2 April 1983

To Pete Roeder

with best wishes,

Al Towle

A PAGURAN ADVENTURE

A PAGURAN ADVENTURE

ALBERT TOWLE

VANTAGE PRESS
New York Washington Atlanta Hollywood

FIRST EDITION

Published by Vantage Press, Inc.
516 West 34th Street, New York, New York 10001

Manufactured in the United States of America
Standard Book Number 533-03635-6

Library of Congress Catalog Card Number 78-54370

DEDICATION

To the memory of my father, Wayne Eugene Towle.

I would like to also thank the several members of my crew (David Feldman, Russell Mabardy, Irwin Thornton, Brad Towle, Nancy Towle, and Karl Wolf) for their faith in *Pagurus* and my ability as captain. Special gratitude must be given to my wife Nancy for her continued support and for her understanding and help in making one of my dreams become a reality.

CONTENTS

PREFACE

This is the story of planning and making a sailing voyage to some interesting and unusual ports of call. The combination of research and pleasure on this expedition taught the author and his students many things about a small boat and the effects of ocean waves, winds, and storms. Many mistakes and misjudgments of boat handling are pointed out so as to serve as valuable information to the beginning cruising person desiring to make a transoceanic passage.

The manuscript has been prepared with the cruising sailor in mind, and also for those persons who have wondered about life aboard a small sailboat during extensive ocean passages. The long days are truly not boring, and there is always much to do and to observe.

Without being too scientific, the author also hopes to provide insights into the flora and fauna of the Galápagos Archipelago. The naturalist may find it interesting to compare my observations to his in other parts of the world.

A PAGURAN ADVENTURE

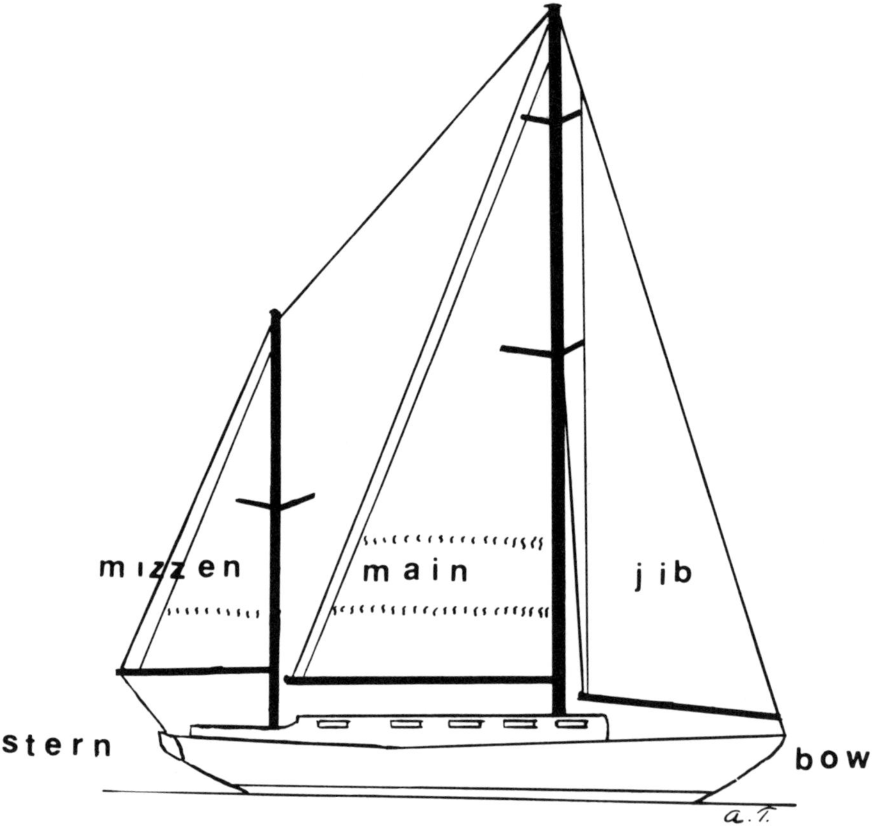

STARBOARD SIDE OF A KETCH

CHAPTER I

WHAT'S A PAGURUS?

How many times have we heard people say, "I've always dreamed of making a voyage in a small boat, but now it is too late"? Many years ago I made a decision that I would never look back and say, "I wish I had." I believe we must always compromise. We have only so much time available to us, and it must be used wisely; this requires planning. For most people, good things don't just happen. They must be thought about, planned carefully, and one must maneuver his activities, assets, and training toward his desired goals.

Sometimes, too, the best laid plans are thwarted. Then, we must reconnoiter and revise our plans to achieve our goals. If an objective is to be achieved, then setbacks are only temporary blocks to be recognized and treated as obstacles. Once these hurdles have been overcome, one can achieve the thrill of a lifetime when a voyage is about to begin, with the words, "Cast off the bowlines."

I would like to tell my story of the voyage of *Pagurus II.* It is not an earthshaking story because small boats are crossing the oceans every day. But it is hoped that in this adventure some of you who will never be able to make such a passage can share in our thrill when our squinting eyes do not deceive us and the small island barely becomes visible after many days of ocean sailing. It is also hoped that I can share with you the adventure and anticipation of planning a long trip. Perhaps I may be able to indicate errors that were made, and you can avoid them. Your voyage may be nothing but smooth sailing!

I would like to begin the evolution of my dream by going back to my childhood. As the fifth child of a family of six boys, we learned to work in the acre of land on which vegetables and fruit were grown. We also had a cow, chickens, ducks, rabbits, and bees. Dad was a postal worker, so our "farming" efforts were an important contribution to our welfare. During the summers, however, we were afforded the opportunity to spend two weeks at my aunt's house in Pacific Grove, California. Here, we learned to swim, fish, and gather abalones for dinner.

At the age of five or six, I can remember walking along the wobbly docks at the wharf in Monterey. The bowsprits of the sailing vessels offered a strange fascination. I touched them and envisioned the bow plunging beneath foamy waves far away from land. If only they could tell the stories of the far ports and strange lands they had visited. Someday, I would have a boat of my own. In the process of education and growing up, however, most childhood dreams are only passing fancies.

World War II came, and then the Korean War. But the vision of white sails carrying me across miles of ocean was not forgotten. I had never even had the opportunity to set foot on a sailboat. At Pensacola, Florida, a small fleet of Stars was maintained for the use of the naval personnel. After a few lessons, I could check out a sailboat and glide through the bayous. This wonderful sailing craft could be beached on the sand for picnic lunches in secluded places away from noise and crowds. One could be alone with his thoughts and be filled with the peace that our beautiful world has to offer.

The next few years were occupied with teaching high school biology and working on a master's degree, and then a doctorate at Stanford. In 1965, however, we had managed to save enough money for down payment on a Herreschoff Modified H-28. Brad, who was twelve at the time, thought of her name—*Pagurus.* As a marine biologist, I thought this name was both appropriate and unique. We christened her *Pagurus.* Hundreds of times since we have been asked, "What's a Pagurus?" I may as well tell you now. *Pagurus* is a

Greek word which means *crab*. It is also the name given to a group of hermit crabs.

Pagurus was a beautiful little boat and taught us much about sailing, navigation, the oceans, winds, weather, and currents. We made two trips between San Francisco and Santa Cruz, and three trips to the Channel Islands. Most weekends were spent in Monterey Bay, and dozens of cruises were taken to Monterey, where we berthed, and to Stillwater Cove, where we anchored.

The adventures we shared as a family had a good unifying effect. Our two daughters and son felt the importance of their contributions to the safety and welfare of all aboard. When they had their friends aboard I could see their faces shine with pride when I announced that it was time for their watches. We were all learning a new respect for discipline that must exist if one is to successfully meet the many challenges that are presented by the ocean, weather, and small boats.

In 1967, I accepted an offer of a position as Professor of Biology at San Francisco State. This meant relocation from Saratoga to somewhere in the Bay Area to decrease the commute distance. We found a house in Millbrae that was to our liking, put a down payment on it, and offered the Saratoga house for sale. As if we didn't have enough to think about, another thing happened that month.

It was a rainy day on 29 May, and I was working at home on some school papers but couldn't seem to concentrate. "Shall we go for a drive?" I asked Nan, my wife. Well, where did we end up? At Sausalito, looking at boats! It was for fun, but once we inspected a 1963 Choey Lee built Winslow Bluewater ketch that was for sale, I said, "This is the boat I want to take on a world voyage." She was forty-two feet and had a Perkins 4-107 diesel engine. We made a deposit and an offer that very day—and spent a sleepless night wondering if we had made a good decision. Two houses and two boats were more than we should have been carrying!

Our offer was accepted, and the survey was conducted on 20 June. She was a good boat. We renamed her *Pagurus II* and had an exciting sail to Santa Cruz.

That summer, I redesigned the wheelhouse and replaced

the twelve-inch wheel with a thirty-six-inch one. This effectively reduced the cockpit size, which I felt was too large for ocean sailing. At the same time, I had an Edson double-purchase worm-gear steering system installed and connected to a Bendix automatic pilot. With our first boat, the northern trips up the coast of California and around Point Conception using a tiller convinced us of the convenience of such a mechanism when only my wife and I were aboard. The installation of a Cole stove in the main saloon added to our comfort when cruising the coast.

Later that year when we sold both *Pagurus I* and our Saratoga house, plans could be laid for a two-year world cruise beginning in 1970. As fate would have it, however, alterations would be made. Our youngest, Nancy Ann, would be graduating from high school, and I thought this would be a good break for her. Brad would have completed two years at a junior college. Linda would have completed college. As the time approached, it became apparent that the United States was still going to be involved in the Vietnam conflict. Brad was doing his duty on an oiler in Westpac. Decision to postpone our plans was necessary.

By 1973, conditions had changed considerably. Linda had married, Brad was a junior at San Francisco State University, Nancy Ann was a student at U. C. Irvine. In January 1973, I made the final decision and asked the family to think about making a year's voyage the following year. We discussed all phases of the trip and the implications on personal involvements. It was then decided, at family council, that each member would let me know his decision by March.

It was a sad day in March when, for various reasons, each member informed me that he should not take a year to make the journey. At times like this, retrenchment is necessary. I took a long walk along the beach looking up at the skyline of San Francisco and thinking about the multitudes of people bustling around at their routine activities. I thought about myself thirty years from now, saying, "I wish I had. . . ." Naturally, financing a voyage, maintaining the home, and keeping two in college would be problems that must be solved. My mind raced as I began to formulate solutions to these major problems. I would be eligible for a sabbatical leave! This would mean a full year at half salary—if various committees were convinced of the academic merit of anticipated research. I could offer research experience to students. As chairman of the Department of Marine Biology at San Francisco State University, I could adequately plan a program and supervise research in marine biology that would contribute to the scientific community.

I could hardly wait to get home and discuss these ideas at the family council. Such a venture is too great an undertaking for a unilateral decision, and each family member would be involved whether he went or stayed. With trepidation, the family agreed it was worth a try, so I began writing a proposal.

I had visited the Galápagos Islands with a group from the University of California one summer, so I knew of some areas that would bear further investigation. Now, I must convince the Ecuadorian government that studies in cooperation with the Charles Darwin Research Station at Puerto Ayora, Santa Cruz Island, would be possible. Many letters were written for approval, pending the award of a sabbatical.

My application for sabbatical for the 1974–75 school year was successful! Now, to select the crew. Many students had heard of the tentative plans and had approached me with enthusiasm and various credentials. After all, I was going to support them for a full year and pay all their expenses except for any personal spending.

Finally I selected two graduate students who had been working a year to meet the requirements but had not yet settled on a project. Dave was married, but his wife gave her blessings, and they began their plans for his absence for a year. Russ was single and had some sailing experience up the coast from San Francisco to Oregon. Both students began to read about the Galápagos and outline their intended research programs.

I felt than an experienced crewman would also be necessary, so I contacted a longtime friend and sailor, Irve Thornton. Irve has a boat and was planning to retire from high school science teaching in 1974. He was excited about the impending voyage and in two months gave me an affirmative answer. Now, he had his arrangements to keep him busy. His daughter, Tia, would be graduating from high school and would begin college in the fall. She was a good manager and a capable, responsible girl. With her approval of her dad's "retirement adventure," she began assuming more tasks of handling finances, so Irve was confident that everything at home would be secure.

With the preliminaries out of the way I could now turn to some of the details involved in planning a year's excursion. I had previously taken two courses in celestial navigation, but that was a long time ago and I hadn't been using it on our coastal trips to the Channel Islands and to Baja California. I enrolled Brad, Russ, and myself in the Winter 1973 course of celestial navigation at the California Academy of Sciences. This course was taught by Captain Todd and Commander Matthews, retired naval officers, and is very thorough, although concentrated. Brad was not planning to make the trip with us, but he was interested, and I thought it would be fun for us to study together. After the course ended on 13 May, Russ and I took the final and received our certificates. This course, however, covered all the theories, and the working of the tables, but it lacked the practical experience of using the sextant. I had a Davis plastic sextant which we used a couple of times, and our computations proved the instrument to be accurate. But it was not until several months later that we were to recognize the significance of a rolling and pitching deck when sighting various celestial bodies! Then, we would realize that working through the HO-211 tables was the least of our problems. After the completion of the course, I worked through several problems with the Air Navigation HO-249, because many friends told me it was easier. However, I didn't find that much time was actually saved by this method, so I printed up 500 worksheets for the HO-211, having decided to use it. Then, I ordered a Heath sextant, which we all found to be very good.

Another important consideration was an adequate medical kit. Two of my sailing friends, Dr. Darrell Sedgwick of Capitola and Dr. Karl Wolf of Manteca helped me assemble an unbelievable list to cover every conceivable contingency. Karl aided me with the procuring of the prescription items. Fortunately the 320 mg. of sealed morphine sulfate, which was kept in a locked cabinet, was returned and signed-off at the end of the voyage. A separate first-aid kit was kept in the dinghy. A list of these supplies may be found in the Appendix.

By 21 February I prepared a *Pagurus II* Newsletter to send to the crew so they would really believe we were going. The newsletter included the planned dates of leaving and route, names and background of the crew, estimated cost of the trip, plans for the projects, selected references of a general nature, a description of *Pagurus II* (including sails, engine, bilge pumps, dinghies, engine accessories, batteries, safety gear), navigational gear, scientific equipment, and a progress report on the medical kit. I believed that such a communication would provide the crew with information to make them more knowledgeable about the boat and preparations. This sharing of the captain's planning did promote interest among faculty, students, and friends at school.

One problem that had plagued me since the beginning of the plans was that of insurance. I wasn't as concerned about the possibility of damage or loss to the boat as much as with the event of injury, or even the loss, of a crew member and friend. The best offer of an insurance policy was $5000, which was prohibitive, so we had a crew meeting and discussed the problem. I drew up a liability release form, and a copy was signed by each crew member and then notarized. My mind was eased having discussed the situation of our joint venture, and I did stress safety. The decision was made. We would go without insurance.

On 20 March, I sent the second newsletter. It contained a progress report, essential items for each person, personal effects, personal equipment, and clothing each should bring. By this time I had been preparing a food stores list, so this was included for the crew's comments and amendments. Medical and dental checkups and a statement of fitness were requested. Information on the Foundation Charles Darwin pour les Galápagos and a chart of the Galápagos Archipelago was sent for information. Included in this newsletter also was the liability release form. (I learned later from my lawyer that liability cannot be signed away.)

Our estimated month of departure had been June or July, but as time drew near, it became obvious that this date was not realistic. There were yet too many things to be done. We all had school responsibilities as well as personal affairs to attend to if we were going to be gone for a year.

On 21 May, *Pagurus II* went back into the water, after her haul-out. When we returned to the berth in Santa Cruz and secured her for the night, we were in a hurry, as I had a meeting in San Francisco. In our haste to depart, we neglected to plug in the shorepower. Realizing this oversight while driving home, we decided that it wouldn't matter, as we were planning to return in two days. This neglect, undoubtedly, saved us from a serious problem. On returning, we learned that several boats in the harbor, especially on our pier, had severe electrolysis. One boat lost both shafts and screws. Another, next to us, had lost the entire screw. It was discovered that one man on our pier had wired a "hot" line to the ground by mistake. I made a dive to inspect all our through-hull fittings, shaft, and screw. All were sound.

The third newsletter went out on 23 May. It was mainly an outline of the items remaining to be done and a work assignment for the crew for the month of June. There was also a request for information to be typed and reproduced for the ship's papers to be used for port entry clearances.

My official sabbatical leave began on 29 May, but the next two weeks were spent with meetings and reading master's theses. Dave and Russ began sanding and varnishing the doghouse on *Pagurus II.* I designed and ordered a new stern pulpit because the Riebandt Vane Steering could not be installed with the existing stern pulpit. My list of spare parts was diminishing, but there were still many items to locate. A complete list of our spare parts and the newsletters can be found in the Appendix.

Each spare part, with a bag of Drierite (water absorbent), was sealed in a small plastic bag. These were then placed in a large plastic bag which was put in a large plastic clothes basket. Fortunately, I found room to stow two such baskets in the engine room. Securely fastened with lanyards, these were accessible when needed—but not easily.

I had a ten-foot Avon rubber boat, but thought we should have another. It was my desire to build chocks on the cabin top and carry a Boston Whaler sailing dinghy. I didn't want to carry gasoline and thought a sailing dinghy would be useful. However, a thorough search failed to locate one, and a new shipment, about to arrive, was to fill orders. I settled on a ten-foot Zodiac with floorboards and oars. This proved to be a durable, useful shoreboat.

On 16 June, we had a dinner for the crew and their families. I showed slides of the Galápagos Islands, and we discussed the impending voyage. We were all getting excited, but there was still so much to do.

The next day I picked up the stern pulpit in Alameda, but two weeks were to pass before it was finally installed. I had another letter to write to the Ecuadorian government for permission to work in the Galápagos, and tried to speed up this tortuous process by going to the Ecuadorian consulate. The situation only became more confusing, so I sent more letters.

The French consulate required that $500 per person be deposited to the Banque de Indochine in Papeete if we were to obtain visas to visit the Marquesas and the Society Islands. Several days were required before international transfers were effected, and it wasn't until 8 July that I finally returned to the French consulate with our passports and the certificates of deposits. We got our visas.

On Sunday, I drove to Monterey to pick up more spare parts, and I visited with Brad, who was enrolled in a summer marine biology course at Hopkins Marine Station. I know he was feeling badly about his decision to stay in school while we were in the South Pacific, but I had to admire his decision, as he was enthused and doing very well.

On 27, 28, and 29 June I was still doing paper work. The crew list was all typed with information in Spanish and in French. These would be used for our entrance into ports and for the *zarpe* (permission to leave) when leaving. Then, I had to make final arrangements with the Smithsonian Institution for carrying certain items to the Darwin Research Station in the Galápagos. Among other things, they needed fifty gallons of isopropyl alcohol. I had no problem stowing the smaller items, but could only carry thirty gallons of alcohol on the deck along the cockpit combing. As you will see later, not all of this arrived.

On 1 July, a diesel mechanic from Alameda came to Santa

Cruz to go over the entire engine with me and to tune it up. At the end of the day's work, he found a leak in the exhaust manifold and said that it should come off and be repaired. Anyone who is familiar with boats knows that just to remove the exhaust stack is not an easy task. It took us two days. There was one nut that defied all efforts to reach it. If we had a one-inch crow's foot, it would be possible. Visits to all shops and hardware stores in Santa Cruz provided negative results. At Moss Landing Marine Manufacturing Company, however, the owner cut an open-end box wrench and welded on a bolt with a hex head so it could be used in combination with another wrench. The next morning we were able to remove the stubborn exhaust manifold and have the rusted spot repaired and the entire manifold sandblasted. Two coats of red-lead paint made it look like new when it was installed.

A truck arrived at our home in Millbrae on 5 July with the food stores. By this time the family room looked like a grocery warehouse, and my den had the appearance of a ship's chandlery. I had planned to stow the canned food in plastic tubs beneath the seats in the main saloon, as they are large areas about four feet wide by six feet long. Hinged doors allow access, and the plastic tubs could be pulled out to select needed items. We began organizing two dozen plastic tubs with canned foods. (Incidentally, this proved to be a satisfactory method of stowage.) Cans that were aboard for over a year were still in good condition; the only cans that rusted were some that were placed in the bilges and became wet with salt water. Since not all the cans would fit in the tubs, many were stowed in the forepeak beneath the two forward bunks. Dave and his wife, Janet, accepted responsibility for stowing the food stores when we brought them down to Santa Cruz. They also made an index card for each type of food, listing the positions of stowage.

The packaged dried foods were placed in plastic bags and the tops tied with string. The starboard hanging locker was converted into a dry stowage pantry for items such as crackers, flours, beans, macaroni, spaghetti, sugar, and spices. This proved to be a dry closet, and we had no problems with spoilage or insects.

Russ and Irve were busy giving the boat a thorough

cleaning below in preparation for the groceries. Dave took the fire extinguishers to a shop to be checked and recharged, if necessary.

On 11 July I had a meeting in San Diego, so I checked the marinas there to see about making reservations for a berth within the month. Everything was booked solid, but I was told that if we just came in they would see what could be done.

The next day Russ and I put all the charts in order, listed them for an inventory, and folded them to fit in the chart drawer. Since the charts were in various sizes, each one had to be considered separately, so the folds would leave a compass rose in each section and the intended routes would not be interrupted by folds. This was accomplished in most cases, and many charts did not require folding at all. The charts were next arranged in order, so we could use them from the top; when we needed the next one, we could place the used one on the bottom.

On 13 July Irve brought the groceries down to Santa Cruz in his truck. Russ and I arrived in time to stow the charts and help carry the food aboard. Dave and Janet had their work cut out for them, and the task appeared impossible to the many envious spectators dreaming of past and future voyages. It was five days before all the groceries were finally catalogued and hidden in proper places. The boot top was almost under the water, but *Pagurus II* remained afloat.

Many friends wanted to see us off and have a bon voyage party, so I set Sunday 21 July 1000 hours for departure. This would give us four days to bring our personal gear aboard and to make a final survey to be certain nothing was forgotten. However, unforeseen events often alter the best laid plans. On 18 July my father, who was eighty-four, had a coronary and was taken to the hospital in Los Gatos. A duty list was made so someone was aboard *Pagurus II* every night while we began the vigil and prayers for dad's recovery. Daily trips from Millbrae to Los Gatos and Santa Cruz were really wearing. Mother was holding up well under the strain, but she preferred to stay at home in Los Gatos.

Dad's recovery was more than satisfactory, and on Friday 2 August I brought him home. He was concerned about having delayed our embarkation! Naturally our primary concern was with his recovery. The prognosis was favorable, so we stocked their house with groceries, and I tentatively reset our departure for Wednesday 6 August.

Dad and mom were fine when we stopped by to see them on the morning of 5 August. I felt comfortable about leaving, so we shopped for perishables, filled the galley, and made the final preparations for leaving. That evening our farewell dinner was held at the Colonial Inn in Santa Cruz. Brad had driven up from Pacific Grove, and his fiancé, Lynn, also joined us. After the emotional strain and months of preparation, the evening was an anticlimax. About 2200 hours we all went down to the Santa Cruz small craft harbor to berth D-13 to say our farewells.

The bow of *Pagurus II* was decorated with floral garlands, and a large printed sign that read:

NO ONE KNOWS WHERE PAGURUS GOES
BUT EVERYONE KNOWS GALÁPAGOS

Aren't boating friends wonderful! And by this time, we thought no one knew when we were going to be able to leave. I had planned that we should get a few hours sleep and leave about 0400 hours. But who could sleep? At 2300 hours I started the engine. At 2310 I kissed my wonderful, understanding wife, shook hands and said good-bye to Brad, who was feeling very sorry for himself, and gave the orders: "Cast off the stern lines." "Cast off the bowlines." It was 2320 when one short blast signaled the departure of *Pagurus II* from her comfortable home port berth at Santa Cruz.

Nan and Nancy Ann bidding us farewell. The sign and flowers by our friends indicate our commitment.

0554 IS MORNING 1754 IS AFTERNOON

CHAPTER II

ROUTINE AT SEA

When leaving Santa Cruz harbor, our spirits rose even higher as the first ocean swell greeted *Pagurus II,* gently lifting her bow. With this thrilling ocean welcome, our adventure was really beginning. The wind, although light, was ample for sailing, but I wanted to give the engine a good shakedown after the tune-up and adjustments. So after passing the Santa Cruz buoy, I set a true course of 163.5°, clicked on the auto-pilot, and drew up a watch list for the next twenty-four hours. But even the off-watch couldn't seem to make contact with Morpheus; the adrenalin of the excitement was over-powering.

At 0330, the RDF (radio direction finder) identified Point Sur three points off our port bow. At 0530, the course was changed to 132.5° to take us to Point Arguello. The next afternoon we were making about 9.2 knots in a 10–15 knot NW wind. Moderate swells were running and a few whitecaps were visible. Later, a light fog enveloped us, and the wind dropped. When trying to sail under these conditions, the sails flop back and forth as the boat rolls, and the condition becomes irritating. Therefore, we furled the mainsail.

Several things occurred at 1915. The engine stopped, Dave became seasick, and Irve had an intestinal upset that demanded his immediate attention. Russ took the wheel while I applied psychoanalysis to the iron monster. I must admit I am not a mechanic, and, although I enjoy most problems, the ones presented by a reluctant engine are not among them. The fact that the engine stopped so suddenly sounded like fuel deficiency, so I inspected the fuel lines, which all appeared to be continuous. Remembering my instruction, and with the aid of the engine manual, I went through the steps of bleeding the fuel lines. The progress really does make sense, but when one doesn't perform this task very often, he forgets the sequence of nuts to loosen. Accompanied by the groans from Dave and the frequent trips to the head by Irve, bleeding the lines was successfully accomplished and the engine was running again in less than an hour. Apparently the problem was bubbles in the fuel line.

At 2100 when I tried to set the autopilot, the clutch wouldn't engage. Well, this problem would have to wait until we reached Dana Point, when we were to have the wind vane installed. At 2125 a course change was was made to 145° as the RDF located Point Arguello 10° off the port bow. Our progress wasn't breaking any records, but we weren't in a hurry because our appointment at Dana Point wasn't until Friday. At 2154 and 2330, the engine repeated its previous conduct by abruptly stopping. The last time it had stopped, I discovered that the filter bowl was loose, so by tightening it I persuaded the engine to run another hour before it stopped again. By now, neither Russ nor I felt as though we could take another hour leaning over a hot, rocking diesel with the fumes saturating our lungs and our stomachs feeling queasy, so we decided to sail in the light, foggy breeze for the remainder of the night and analyze the situation by daylight.

In the morning, with Irve at the helm and Dave lying very quietly in the bunk, Russ and I began inspecting the fuel lines once again. Starting at each fuel tank, we checked the valves, the lines for leaks, and the primary and secondary fuel filters. When we reached the bowl filter on the engine, I removed it and noticed an irregularity in the gasket. After a discussion, the decision was to cut a new gasket for the bowl filter. The new gasket was put in place, and when the filter bowl was assembled, it was filled and the lines bled once more. This certainly must have been the problem because the faithful Perkins 4-107 never gave us any problems for the remainder of the voyage.

By 1330 that afternoon we were on the SE end of Santa Rosa Island, and heading towards Los Angeles harbor. The next afternoon, 8 August, we tied up to the fuel dock at Dana Point. The harbormaster assigned *Pagurus II* to berth 10 and told us we could only stay for five days (but, of course, if unforeseen problems developed, the time could be extended).

It was good to have the shakedown leg of our journey completed, and Dana Point was warm. Dave went for a walk and felt better. At dinner that evening, however, he informed me that he thought it best not to continue on the expedition. It was with mixed emotion that I accepted his decision, and I still believe he would have overcome the malady of the sea. I am sure he was also missing Janet, because his eyes took on new life when she came down the next day. She also brought another emergency life raft for us to carry on the trip.

Nan and Lynn drove down to Dana Point the next evening. Lynn was going to her home in San Diego and Nan was going to stay with us to enjoy the short cruise to San Diego. I proudly showed her the neatly installed Riebandt Vane Steering. Everything fit perfectly and *Pagurus II* now really looked ready for a voyage.

The next few days were spent in reorganizing and restowing things below while waiting for a technician from Harbor Radio Service to locate the electronic problem in the autopilot. Then, we had to wait for a capacitor to be airmailed from the factory. I was glad that the shakedown cruise had been planned to identify and correct these problems before leaving the United States.

On the evening of 15 August, we invited Kramer Rohfleisch and his wife from the *Solveig* aboard for cocktails. They were on their way to the Channel Islands for a month and asked if we would like to use their berth at the Southwestern Yacht Club when we called at San Diego. Naturally, he received an overwhelming positive answer, so he wrote a letter to Commodore James, thus resolving the concern of locating a San Diego berth.

On 16 August *Pagurus II* left the Dana Point buoy at 0740 and took up a course for Point Loma. It was a beautiful sunny cloudless morning. We tested the wind vane, then the autopilot, and took noon sights for practicing our navigational skills. We thought about Dave as we contrasted this short trip with the last one. Nan was enjoying the pleasant Southern California weather by sitting on the cabin top in the warm sunshine.

At 1800 when we came into Southwestern Yacht Club, there was a boat in berth C-49, so we went into A-47. I couldn't locate the port captain, so one of the members suggested we remain in A-47 until morning. Nan fixed beef stroganoff and we had a relaxing evening thinking about how perfect the day had been.

The next morning we moved to berth C-49, next to the beautiful *Quiscilla,* a Winslow-designed ketch built in 1924. The Hootons lived aboard, and she was very comfortable, as well as being maintained for a voyage at a moment's decision. Of all the people we visited, the members of Southwestern Yacht Club will always be fondly remembered for their hospitality. Dave Hooton loaned us his car when he and his wife left for a month's vacation. This amenity eased the burden of shopping and effecting transfers to and from the San Diego airport.

Brad had called from Pacific Grove, and when he informed me that he would like to take the year off from school, I was very happy. Since he was a marine biology major, I arranged for him to work on a project to earn research units during the anticipated stay in the Galápagos Archipelago. Therefore, I was certain the year's experience would be of more than equivalent value to an academic year at the university.

Our departure date was delayed a week while waiting for Brad, but the time was utilized by making baggywrinkle, testing methods of growing alfalfa sprouts and mung beans, and we seemed to have last minute shopping. Baggywrinkle is material attached to the shrouds to prevent chafing of the sails from rubbing on the stainless steel shrouds. Also, since Dave had performed the task of stowing all the food aboard, we went over the three-by-five cards which he had typed (indicating the location of the various items) to be certain we could find specific foods when desired.

On Thursday 22 August I took Nan to the airport to say good-bye for ten months. This was to be our longest separation since we were teenagers, and I know that neither of us could put our innermost feelings into words. At such times, quiet understanding is a form of communication that passes between us. It seemed so this day, as we quietly sat in the terminal waiting for the San Francisco flight to be announced.

By Saturday, the baggywrinkle was installed on the mizzen shrouds and *Pagurus II* was finally ready for a passage. Everything was in order when Brad arrived on Sunday. We celebrated the eve of our departure with champagne and a steak dinner aboard.

The next morning we were ready to leave the world of crowded freeways, noisy airports, and smoggy air. At this moment, one wonders if everything has really been done in anticipation of all contingencies. After loading 350 pounds of ice, and several cases of soft drinks, we left the San Diego fuel pier at 0930 on Monday 26 August. Outside the channel, the mizzen and jib were hauled. Then, as the main was being hauled, a shackle on the first grommet came loose and ripped about a foot of the mainsail near the peak. Damn! After considering returning for repair, I decided that this was to be our first "emergency" of the voyage and that we would face others, so here was a good place to demonstrate our independence. Russ and I spent most of the afternoon sitting on the foredeck, hand-sewing the tear as we clipped along at seven to eight knots under the mizzen and jib. It was a beautiful afternoon, and we could see the Coronados as the mainland receded.

During the night the force of the NW wind increased, so the main was furled, and remained furled for four days. The jib and mizzen were sufficient to take us about 125 miles each day, and we were comfortable as we began to develop a shipboard routine. Watches were staggered so we could take turns preparing the evening meals, which became the highlights of the days. We were still using our fresh supplies and hadn't yet taxed our innovative culinary skills with the challenges offered by the stored canned goods.

After dinner, the evening stars provided us with a fix to chart our progress on the universal plotting sheet. Then, coffee or tea in the cockpit offered company for the first evening watch. Around 2300, when Morpheus called taps for the off-watch crew, I found that learning to play the harmonica was an enjoyable pastime. At first I wondered that the crew didn't complain about the "sounds" issuing from the cockpit, but after listening to others amuse themselves in the same manner, I realized that hearing the harmonica was very reassuring to those below. In fact, very often I suddenly awoke when the watch stopped playing to devote his full attention to another matter. When one has become acclimated to life at sea on a small boat, sounds become very important, and the mind sorts out the steady, rhythmic, comfortable ones from minor irregular interruptions that may indicate trouble.

Mornings at sea, these first few days, were beautiful. The gradual lightening of the sky at dawn, and then the sun showing its upper limb above the horizon offered the promise of another day's travel, carrying us closer to the Enchanted Isles. This was a moment I especially enjoyed when I was on watch. When everything was going right, the wind vane would hold *Pagurus II* on her course, and I could write in my journal before the 0600 relief would come up on deck with two cups of steaming coffee. At times like this there was a feeling of complete harmony with one's place in the environment. The freshness of the clear, pure air with a slight salty smell, the swishing sound of the bow cutting its way through the clear blue water added to the enchantment of merely being alive. Sudden realization that surrounding conditions have provided a setting for this attitude of inner peace was a recurring topic for my journal during our voyage. If only we could recall these conditions and these thoughts every morning when we awake, no matter where we are, we could all lead happier lives. Certainly, every voyager has had similar experiences when he reconfirms his affiliation with God. Difficulty arises when he returns to daily routine and bogs down with boredom, forgetting that he is capable of being aware of natural beauty.

Often, in the early morning and with such thoughts in mind, a new awakening of an immediate nature would occur as the smell of bacon began to arise from the galley. We were all ready for another day.

On the fifth night out, the wind dropped as we ghosted along at three knots. We weren't able to sight the evening stars, and the cloud cover also prohibited a morning fix. With nature's disappointment, however, came squalls, and we were able to take freshwater showers. It was good to rid ourselves of the stickiness of daily saltwater bucket-baths.

During Irve's watch on the sixth night out of San Diego, he called us to see the lights of a ship. We had been traveling without lights to conserve the batteries, so now the running lights were switched on. The ship was a tanker, and she stopped about a quarter of a mile away. As her deck lights blinked on and off, I listened on the radio, but received nothing. Then, I put on our spreader lights to show the sails. After twenty minutes, the tanker blinked her lights and steamed off. It was comforting to know that had we been in need of aid, a flare would have, undoubtedly, brought assistance.

Warm weather and rain squalls continued, but scattered clouds on the morning of 1 September allowed four star shots that gave credulity to our D.R. (dead reckoning) position of latitude 22° 41.2′N and longitude 114° 30.0′W. Although our desired course was 115°, we could only average a course of 215°, and the wind was dying. For the following day we were becalmed. The log and journal indicate how life was aboard *Pagurus II*.

After being becalmed in almost glassy seas most of the night, Russ and I were sitting on the port walkway at 0730, noting how slowly the air bubbles were flowing past. We were soon aware that many "air bubbles" appeared to suddenly dart around in an irregular fashion, leaving a wake on the calm water. Closer inspection revealed them to be an iridescent blue-green. After several attempts, we managed to capture one of these mysterious objects, and examination with a hand lens showed it to be a marine insect, *Halobates*. These insects are highly specialized for life on the surface of water. Although their freshwater relatives have wings, the marine forms are wingless. It was interesting to observe these small insects on the ocean so far from land.

We also observed that a beautiful fish had joined us. From the surface it looked torpedo-shaped and a beautiful iridescent blue in color. It appeared to be almost three feet long. (We were to learn later that this dolphin fish, or dorado, of the family Coryphaemidae, was to provide a welcome change from canned food.)

Since we weren't moving at all, even with all sails up, we went swimming in the beautiful clear blue water of the Pacific. After our swim, we noticed clouds approaching, so we prepared for a refreshing freshwater shower. Then, when the sun came out, Irve made breakfast of oatmeal and raisins, in an interesting manner. To conserve fresh water, he used salt water. This proved to be just a little too salty, but we managed to convey appreciation as our breakfast was slowly consumed.

A little bird joined us; it was a swallow, and I wondered if it would ever find its way back to land. (It stayed with us for a day and a half.)

For dinner, we had ham, potatoes and onions, mung bean sprout salad with kidney beans, and tomatoes. Then, coffee was enjoyed in the cockpit as the wind pushed us along at six knots. Our stereo played one of Mario Lanza's tapes, and we all agreed it was most enjoyable here in our "small world."

Dinner over, the watch list was posted, and we remained in the cockpit for star-recognition review. We hoped the wind would hold, because being in irons for nearly two days had been difficult—even though we were able to wash clothes in fresh water and get them dried in the sun.

At 2020 we saw a satellite pass through the sky from west to east. This "omen" seemed to indicate to the off-watch that it was time to retire. A lone sentinel then took command.

On 3 September, the wind picked up and we traveled eight knots nearly all day. Live flying fish littered the deck in the morning. Our little bird was dead and found on the cockpit grating. Brad and Irve took the helm for the entire day while Russ and I worked out some navigational problems. We had been getting some good sun lines, but hadn't been able to get a good star fix, so our longitude was assumed from the sun shots and the D.R., keeping track of course changes.

On this particular afternoon we passed through a long-lasting heavy rain squall. Irve and Brad stayed out to bathe. Russ took a long nap, and I enjoyed reading and listening to the pelting rain on the cabin top.

Evening star sights did not give us a good fix. It seemed as though a cloud would cover the sighted celestial body just at the crucial moment when the sextant was bringing it to the horizon. "Perhaps we'll have better results in the morning," I commented.

At 0500 Brad woke me and we got a good four-star fix. With confidence, we now knew where we were: latitude 21° 22.5'N, longitude 113° 47.0'W. This was about 600 miles west of the Mexican Coast and about 400 miles south of Cabo San Lucas, Baja California.

The wind was from the SE, so we were required to make long tacks, but our speed was only three to three and one-half knots. Our ice vanished on this day, so I cleaned out the refrigerator and we stowed the soft drinks in the cockpit ice compartment. We still had fresh corned beef, cabbage, tomatoes, and carrots. These items kept surprisingly well.

The weather, although not ideal for making headway, had really been beautiful. The sea was a cobalt blue, reflecting the sky, dotted with cumulus clouds. "Tonight, after supper," it was decided, "we will play harmonicas and talk until it is time to take the evening star sights." Everyone commented on the peacefulness and "forced" slow pace to which we were really becoming accustomed. We hadn't seen a ship for four days, and it was truly a world of our own.

On 5 September the winds were very light and still from the wrong quadrant, so we were ghosting along between two and three knots. The morning was warm and beautiful, but during the early morning watch there had been showers. The morning star fix, using Capella, Betelgeuse, Rigel, and Sirius, showed that we had made ninety-two miles during the last twenty-four hours. However, we were heading east and wanted to go east by south. I guessed that we were breaking into the "cruising life," however, because we weren't so concerned about the extra days it would take. During the next twenty-four hours we only made thirty-nine miles!

When averaging our speed since leaving San Diego, we found that during the first four days we made 120 miles each day, but for the last week our average was only 30 miles per day. We had been in doldrums with scattered, slow-moving squalls and showers, and it was very hot. Now, on 8 September, we were on a broad reach traveling on a true course of 115° at six knots.

By 9 September, we had reached a latitude of 19° 48.0′N and a longitude of 109° 16.5′W. Although a speed record wasn't being made, Pagurus II was only twenty miles east of the course line which I had plotted when we were in San Diego. We should be able to find the Cocos Islands! We all seemed to be well adjusted to life at sea, and duties were performed efficiently. Our spare time was utilized for reading, backgammon, observing the feeding storm petrels, and planning meals.

At 0630 one morning, a brown booby was thwarted in its first attempts to land atop the mizzenmast. Just as the approach was made, the mast swung toward the bird and the landing was missed. The second time, the mast swung away and, the bird hit the mizzen sail. Success was achieved on the third try, and the bird clung to the mast top with its webbed feet, swaying its body to maintain balance on the arcing perch. Irve didn't like that bird up there and tried to throw things at it, but the booby wouldn't be discouraged. Three times that day we were required to swab the cockpit under the resting bird. All day, the bird's "buddies" tried to gain the perch, but the possessor maintained its claim. Night approached, as did Irve's watch, and he became very determined that the bird had to go. A swab was half-hitched to the mizzen signal halyard and hoisted to the level of the bird. By jerking on the halyard, the swab could be made to tap the bird on the breast. Its annoyance was demonstrated by a squawk, as it pecked at the swab. After an hour the bird was finally dislodged. It made several passes attempting to regain its perch, but perhaps in the dimness of twilight, depth perception was "off," because it finally left us, much to Irve's expressed joy.

Brad, Russ, and I were all getting quite brown. Irve burned his knees and stomach, so he was staying below most of the day. However, it did get stuffy below, and I found it more pleasant on deck in the shade of the sail. Refrigeration would have been a big help in these temperatures, but we seemed to be managing without it. The margarine was liquid, as it never seemed to cool off enough for it to solidify; it was good that we had contained it in plastic bags. We also had some liquid margarine, but wanted to use up the "solid" type first. The eggs, which had been rubbed with mineral oil, were keeping very well. Our sleepings bags became useless now, and the vinyl pads of the bunks were not very comfortable.

On 12 September, we spent another day of ghosting along at one-quarter to one-half knot, with the sails flapping. I ran the engine for an hour to charge the batteries and see what it felt like to move along. The breeze we created felt good, but disappointment was obvious when I shut down the engine. The September pilot charts didn't indicate these fluky winds in this area at this time of year.

Other forms of life weren't very numerous to break the monotony of the slow travel. On the previous day we saw three brown boobies, a few flying fish, two dolphin fish, and a school of small slate-gray fish about sixteen centimeters long. On this day we passed through thousands of small coelenterate medusae. The bells of thse jellyfish were about five centimeters in diameter and the feeding tentacles were long, flattened, and brownish in color.

A terrific thunderstorm with bright lightning streaks and heavy rain gave Brad a good freshwater shower about midnight. It didn't bring steady winds, however, but an overcast sky followed and prevented us from getting a star fix in the morning.

The night watches on 13 September were entertained by schools of porpoises whose antics could be followed by the luminescent streaks marking their wakes. Our speed during the night was about one knot.

The next day I scheduled a "field day." Three lashings on grommets of the mainsail had parted, so these were resewn and all sail grommet lashings were checked and reinforced where necessary. The decks were washed and teak oil was applied to the rails. The change of pace and required duties were really welcomed by all aboard.

By the afternoon the SE breeze became steady, and we were making five knots. The scattered clouds of evening promised the opportunity to get a good star fix, and the setting sun created a spectacular evening. Irve and Russ had made dinner of tortillas, chili beans, chopped cabbage and onions, and a salad of bean sprouts from Russ's "garden." With catsup, parmesan cheese, and Tabasco sauce, we combined these into delicious tacos. Red wine complemented this fare, and the wind vane held *Pagurus II* on course as a faithful helmsman.

Then, the wind became so variable for the next few days that the wind vane was not helpful. The sky was overcast, and we experienced several squalls both day and night. It seemed strange to observe bits of bark and twigs floating in the water. A brown booby joined us and wouldn't be discouraged. His perch was the stern pulpit. Whenever other boobies would approach, it would squawk a warning that this was his territory. I really hoped he would keep the others away, because one is enough! As long as his tail faced outboard, it was all right, but once when he was warning others away, we had to wash down the poop deck. When Irve woke up, the booby's flight time arrived!

Strong variable winds accompanied the rainy squalls, which became more numerous, so the mainsail was furled even though the intermittent periods were almost without any wind.

By the night of 19 September, the winds increased and steadied from the NW, so we were again making seven knots. During the night the rigging glowed with St. Elmo's fire, its eerie greenish glow intriguing. At 0820 the next morning, we passed one of the most beautiful and cleanest looking freighters I've ever seen. She was the *Blue Master* from Oslo, and must have come up from Panama.

We had made eighty-two miles in the last twenty-four hours, which wasn't too bad, considering the storms. The freshwater showers during the days were refreshing, but during the night the storms and lightning were not as pleasant. We were now running a steady course and the wind vane was again holding well. When these conditions prevailed, I awoke at 0445 and made coffee for Irve (who had been on watch for two hours) and me. Then, I would wake Russ to record the morning star sights. After this was done, Irve would go to bed, Russ would go to bed, and I would have two hours of watch left. If conditions were favorable, I could set the steering vane and go below to chart the night's passage to get our D.R. position. Then, taking the nautical almanac, Ageton, paper, and pencil, I would return to the cockpit with another cup of coffee. If the vane was still holding a good, steady course, I could then do the computations for the morning fix and then go below to plot and record the position in the "star log." Then, by the time the fix was transferred from the universal plotting sheet to the chart, it would be nearly 0700 and time for Russ's watch. If everything was running smoothly, I didn't care if the crew slept in because I then had quiet moments to write in the journal and to reflect.

Since this part of the trip had taken so long, my main concern was about everyone at home worrying. However, Irve had read in a back issue of *Sea* magazine that one voyager took sixty-one days to get from Panama to the Marquesas!

What a beautiful morning it was. But it was to be the calm before the storm. The next learning experience was to be one of the most traumatic of our voyage, perhaps because it was the first experience of heavy weather or because the doldrums had lulled us into a sense of calm security.

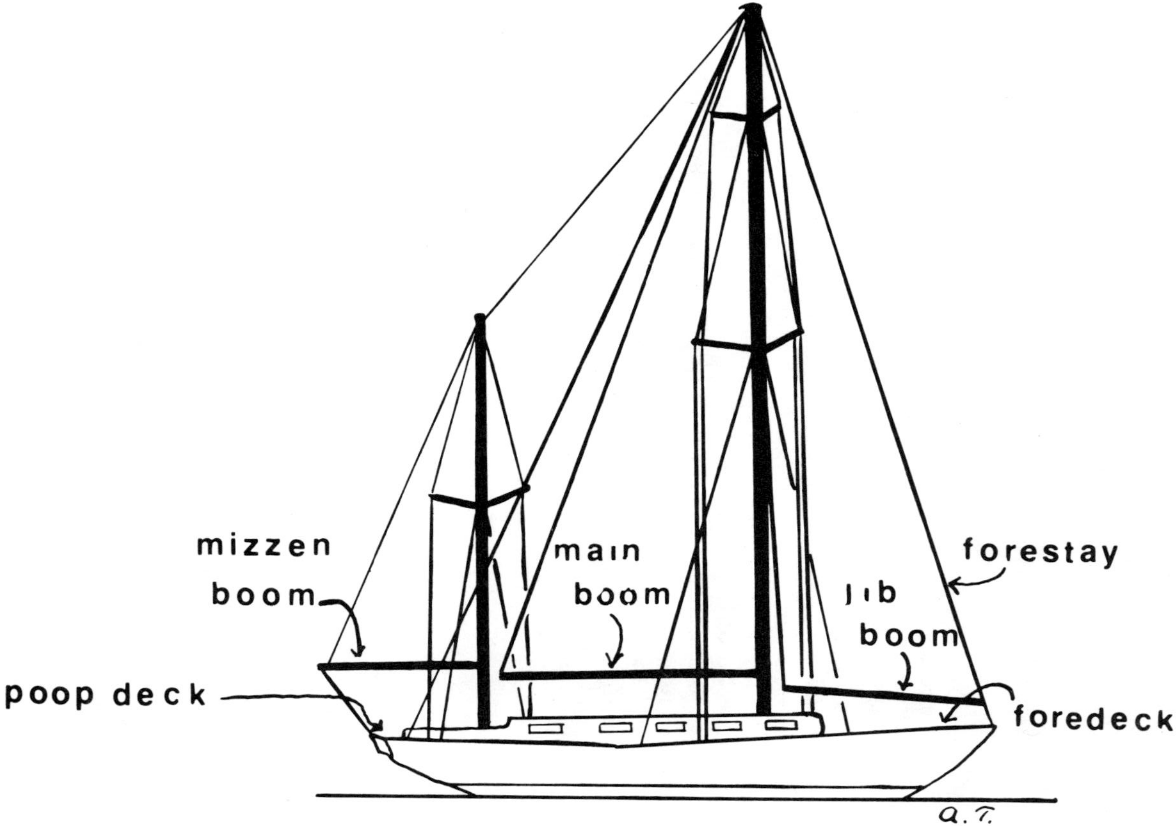
mizzen
boom
main
boom
jib
boom
forestay
poop deck
foredeck
a.T.

KETCH

CHAPTER III

ACAPULCO

On 20 September the winds were so variable in direction and force that I cut the watches down to two hours. On the various watch periods during the day our speed averaged from 4.3 knots to as high as 7.8 knots. Through our dinner hour the wind vane held, and even though our speed had increased to nine knots, we were on a broad reach and the angle of heel was comfortable. We sat in the cockpit sipping mocha-coffee, as I began to analyze the increasing velocity of the wind. It was then decided to have two people stand four-hour watches through the night rather than each of us standing two hours at the helm.

Irve and I were on watch when, at 2200, I said, "Well, the wind seems steady although it changed from NW to SW, so you go below and I'll wake Russ at 2300." I had planned to stay with Russ until 0100, and if conditions remained unchanged, he could take the helm alone until 0300 when he would awaken Brad. Then, I thought that he could keep Brad company for the first hour. Setting up such a staggered watch, we could all get some rest through the night, and yet be available if needed.

Russ came out a little before 2300 because the wind was really whipping up, and it had begun to rain—not little drops, but buckets! Since it had been very warm, we were wearing only shorts, but soon we were getting cold. In our past experiences, these rains lasted for ten or fifteen minutes, and we would pass into blue sky and stars again. However, this was obviously different. An hour and a half later it was still pouring. Intermittently, Brad and Irve would come to the hatch and ask what they could do—but nothing seemed imminently necessary.

The next thing we knew, Brad had made pancakes (of all things!) and Irve handed Russ and me each a bowl of pancakes with syrup. Well, here we were sitting in a thunderstorm, with lightning so close we could feel the heat, eating pancakes in the drenching rain in the Pacific Ocean somewhere 250 miles south of Acapulco! We finished this fare and laughed at the incongruity of the situation and at our good fortune in averaging nine knots toward the Cocos when the outhaul on the main snapped. (In anticipation of improving weather conditions, I wanted to let the crew rest as much as possible so I hadn't called for furling of the mainsail. This was the first mistake.) I put on the spreader lights, started the engine (which brought Brad and Irve to the hatch immediately) and yelled, "All hands on deck, we have to douse the main." By this time the seas were really high and the winds exceeded sixty-five knots.

I released the jib sheet to let it out a little before Brad, Russ, and I furled the main. Irve manned the helm to hold her into weather, which was almost impossible—even with the engine. The mainsail was no sooner furled when I saw, to my horror, the jib boom rise up so that it became nearly parallel to the forestay. Apparently, in the urgency of the moment, I hadn't properly secured the jib sheet before helping furl the mainsail (second mistake).

When the jib boom snapped back into position, three jib snaps had become unsewn, thus a decision to furl the jib until repair could be made in the light of day. At this point the mizzen was reefed to act as a vane, and we broke out two newly purchased plastic sea anchors. As soon as they were deployed, they tore into shreds, leaving trailing lines in the water. Fortunately, we had another old canvas sea anchor with an iron ring. Irve brought this one as a spare. It held, giving us a more secure position in the heavy seas. Now, at least, we could take turns getting dried off, warm, and taking intermittent rests.

At 0530 Brad and I went to the bow. With howling wind and breaking seas, the two of us managed to resew the snaps onto the grommets of the jib. However, since our hove-to position was as comfortable as possible, the decision was made to wait for a while before hoisting the jib. We then checked all the slides on the main and reinforced the lashings where they appeared weak. Irve and Russ were manning the cockpit.

We took turns drying off below, resting, and then donning foul-weather gear before going topside again. At 0900 the jib and mizzen were hoisted and the sea anchor hauled in. We were finally underway again at seven knots. I made coffee, and Brad and Russ took the watch until Irve and I relieved them at 1100. Although it was still pelting rain with heavy winds and seas, the motion below was much better under way than when we were hove-to. This storm was certainly more extensive than any we had experienced so far, and the winds had exceeded 100 knots. We also lost our starboard deckload of ten gallons of diesel and ten gallons of alcohol (which I was carrying to the Darwin Research Station in the Galápagos). Certainly, the third mistake was not securing the deckload properly—although we thought we had. I began making notes of these things we must check and do when the storm abated. I noticed three slides on the mizzen which would need reinforcing.

Fortunately the mizzen and the main booms were properly vanged. This had been done by running lines from cleats on the deck to the boom, so as to limit the distance the booms could swing out. During the emergencies of the night we weren't able to secure the main boom in its proper position, but we secured it on the port side by the topping lift, vang, and sheet lines.

During the next night (21 September) double watches were again set up because the end of this foul weather was still not in sight. At 2400, Irve and I were relieved by Brad and Russ. It was so good to get warm and into a prone position for a little rest. I think I was almost asleep when there was a sharp sound and Brad shouted, "The forestay parted." In a flash we were on deck where we could see the jib boom banging on the cabin top as it swung wildly with the wind and the pitching, rolling boat. Irve took the helm to bring her into weather while I started the engine. Brad, Russ, and I worked on securing the job boom and untangling the fouled sheet, as the decks rocked, rolled, and pitched in an unpredictable manner. The tension placed on the forestay by the two (now unstable) masts and the jib boom added to our difficulty. Luckily the turnbuckle was still attached to the bolt in the bow—only by one turn. We tried several ways to pull and tug to approximate the forestay bolt to the opening in the turnbuckle, but in no way were we successful. In the dim light offered by our spreader lights cutting through the rain of the black night, our task seemed impossible. Finally I said, "We will just have to unbolt the jib boom from the forestay." Since we were afraid of losing the forestay in the air, Brad suggested that we should tie it down. Then, I got an idea of tying the forestay and running the line through the block of the jib sheet and over to the power winch. When I started this procedure, I rove the line through the bow anchor roller instead of the block of the jib sheet; it was a more secure route for the line. Fortunately, the wind had abated somewhat by this time (about fifty knots, the lightest it had been for thirty-six hours), but the bow was still pitching wildly. We would be way up in the air at one moment, and then the deck would fall away and green water would envelop us to our waists. We removed the jib boom and pulled down the forestay, but getting the bolt to match the turnbuckle took a long time and a lot of yelling. We did it at last!

After tightening the turnbuckle and replacing the jib boom, we had to bend on the jib (which had been removed). Next, when we tried to raise the jib, we soon found that the jib halyard had, somehow, become fouled aloft, and we couldn't free it. We would have to wait for the light of day. Irve and I went below for twenty minutes before we were scheduled to go on watch. It was then that I realized that we had done all this work, and I didn't have on a stitch of clothing!

Sitting at home and examining these problems may seem simple, but in the excitement of the moment of emergency, situations are quite different. Likewise, the light of dawn brings with it a certain calmness, even in rough seas. The jib halyard was freed, and the jib raised. We picked up speed and, with the jib and mizzen, were recording nine knots; the storm still raged with gusts exceeding sixty-five knots.

In the predawn hours Brad had noticed a creaking of the mizzenmast and reported it to me when I came up to take the watch. Inspection revealed that the mizzenmast (hollow spruce) was split at the seam below the bridge deck. It was stepped on the floor of the cockpit, and with each pitch of the boat, it make a cracking noise. Tightening the two backstays helped a little, but by 1000 on 22 September (with the storm continuing), I doused the mizzen sail and we recorded seven knots on the jib alone. How to repair this at sea was a puzzle, and I mulled over several possibilities, but to go on seemed to be asking for trouble. At 1230 on 23 September, we altered course to 347° for Acapulco, where I had hopes of effecting repairs. The estimated distance was 120 miles.

I had no charts of the Acapulco area except the large planning chart, and with only the jib to help pull us along, the engine was started at 1525. Although the rains had been continuous since the previous night, the seas had considerably flattened, so I stationed only one man on watch so we could get some much needed sleep.

Our spirits were raised that night with the culinary efforts of Brad and Russ. We had a tamale pie. The next morning at 0730 Acapulco radio stations were picked up on the RDF. Bearing readings showed Acapulco to bear 300° and not 347°, so the course was altered. At the time of our course change, a Japan Lines freighter, *Japan Ambassador,* crossed our bow. At least we weren't alone. The worst of the storm was over, and the sun began to peek out through the heavy clouds.

Clean-up began; Brad was on the helm. Irve and Russ washed out the smelly clothes in salt water and rinsed them in fresh water which had been collected in the dinghy. Russ opened the ports and took up the carpets. I was certain we would find Acapulco this night and make port in the morning. The galley required much attention. On the night the storm hit and Brad had made pancakes, he had "saved" some batter for breakfast. This was three days before. Also, a box of Tide got wet during the storm, opened, and scattered over the galley area. Since the icebox storage area had been open to the air, much of this mess had accumulated here. For three days the mixture of batter, Tide, and water had been oozing back and forth, much to the apparent "enjoyment" of bacteria and mold. This amorphous slimy mixture was found everywhere. Once, when soup had been made, violent lurching threw some of the contents over the stove top and in the galley area also. Matches were damp and wouldn't light. Improperly wrapped packages of pudding, spaghetti, cookies, and crackers were discarded. But Acapulco was near, and we could replenish some of these provisions. With tired but happy hearts, the galley was cleaned, the main saloon straightened, and *Pagurus II* once again began to resemble a proud little vessel.

It wasn't until 24 September that we began to see several ships and even small fishing vessels. We entered Baca Grande at 0500, but it wasn't until 1230 that we dropped anchor near the commercial dock and raised our quarantine flag.

Two days later we learned that we had encountered Hurricane Fifi after it had ripped through Guatamala and then swung back up to Mazatlán.

CHAPTER IV

BAQUERIZO MORENO
(Wreck Bay, San Cristobal)

Bahia del Acapulco is a very scenic and well protected harbor. The green mountains, scattered beaches, coconut palms, and picturesque houses offered a welcome contrast to the hostile storm we had just experienced. As in most harbors of the world, the water surrounding our anchorage did not tempt us enough for swimming, in spite of the warmth of the sun on the decks.

With the quarantine flag flying we expected the capitán de puerto momentarily. Since we wanted to represent the boating people in the best perspective, everything was made shipshape in jig time. We waited. At 1600, I realized they wouldn't be out until the next day. Well, we couldn't go ashore, but the awesome stillness of the peaceful anchorage was such a contrast to the month's motion in passage that I wondered if we would ever dare to set out on the sea again.

The next morning a customs man and a health inspector were brought out by a boat. The cost was twelve dollars for immigration, eight dollars for a certification of health, four dollars for the boat, and two packages of American cigarettes. They also pointed out a boat yard which consisted of a sandy beach surrounded by tall coconut palms. After a "*gracias*" the quarantine flag was lowered, the anchor weighed, and we moved over near the boat yard and dropped the anchor. Irve and I then set out in the dinghy to find Señor Pepe Vega. He

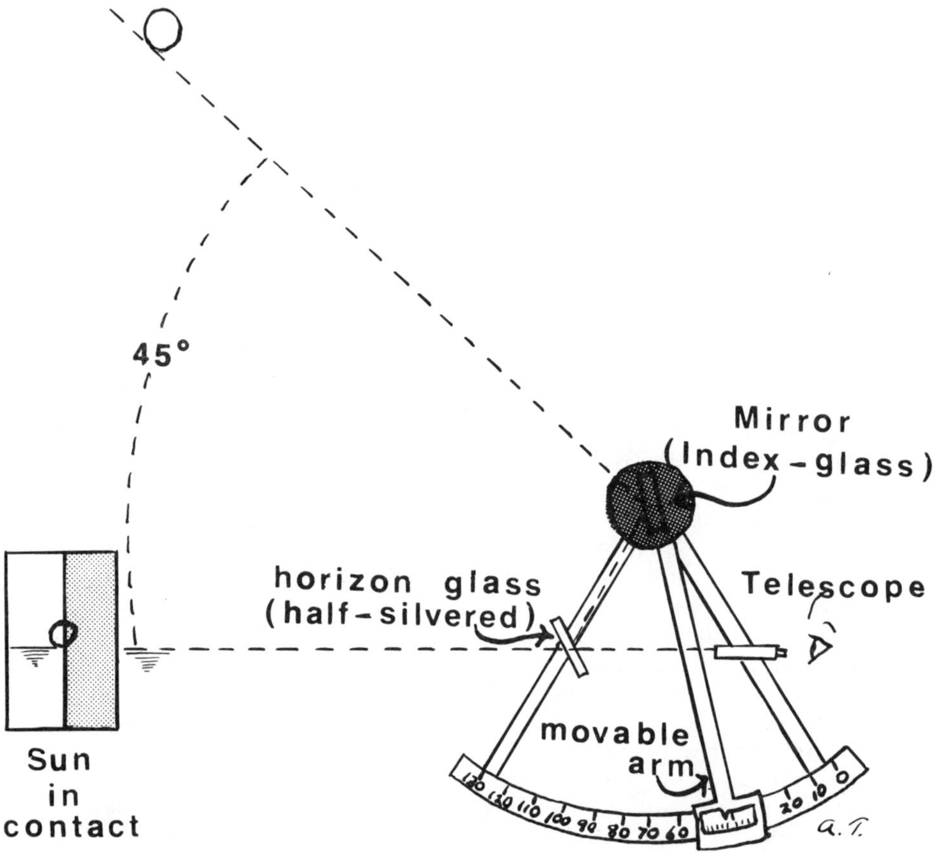
45°
Mirror
(Index-glass)
horizon glass
(half-silvered)
Telescope
movable
arm
Sun
in
contact
120 130 110 100 90 80 70 60
20 10 0
a. T.

THE SEXTANT

wasn't at the yard, but we were shown the way up steep steps to his villa on the cliffs.

Señor Pepe was in Mexico City, but Señor Martinez would be back shortly, and he would come out to our *"barco."* Brad and Russ were very anxious to go ashore, so when Irve and I returned to *Pagurus II,* they were waiting. I spent the afternoon sewing sail grommets to the slides and fixing the outhaul on the mainsail while Irve sanded and applied teak oil to the rails.

When Brad and Russ returned, they announced that the yacht club was plush and the fees were twenty dollars per day to tie up there. We could also obtain our diesel fuel at *Acapulco Club del Yate.* I then decided that once our mizzenmast was properly fixed, we could go to the club for a few days, enjoy freshwater showers, and do some laundry. Also, it appeared as though the main boom should be removed and reglued in some areas. If we could purchase some large clamps, this could better be accomplished on a dock than at anchor. It was quite obvious, to me, that we would be here for more than a week.

Two days later we saw Señor Martinez! When I explained that I would like to find a machinist to make two straps of *"acero inoxidable"* he shook his head and explained that it would probably be iron, but I might (if lucky) obtain bronze. Then, he kindly drove me to Roberto's machine shop. After firm insistence on my part, Roberto reexamined my diagrams and measurements before agreeing to locate some bronze straps and eight large bronze nuts and bolts. He would have it ready in two days and the cost would be 750 pesos (sixty dollars). I returned to *"Barco Pagurus"* happily, as I planned an afternoon and evening ashore with the crew.

Perhaps one of the joys of traveling by small boat is that one can be in a foreign port, but always return to familiar surroundings. Since most people travel to foreign lands by air, their area of solitude becomes a strange hotel room. Here, we were in Acapulco, and yet we were "home" and didn't have suitcases through which to search for proper attire. The feeling of being here, yet being able to leave when we wished and not tied to a schedule gave us all a carefree feeling which was exhilarating. None of us had ever been to Acapulco before.

We took a bus into Acapulco and then walked to the *mercado*. We talked to people, even though our Spanish left much to be desired. We investigated clothing shops, hardware stores, bookstores, and grocery stores before having supper at a good restaurant. Then we stopped at the telephone exchange to surprise everyone at home by unexpected telephone calls. Nan informed me that my request for investigative studies in the Galápagos had been approved by the Ecuadorian government, and that Darwin Research Station had been notified to await our arrival. Although I had been confident that approval would be given, the final acknowledgement was certainly reassuring. Now, it remained to insure the seaworthiness of our vessel before leaving Acapulco.

My spirits were dampened that evening, however, by a discussion with Irve as we were returning to the boat. I knew that Irve was concerned about being away from home for such an extended period. He had decided, that evening, to return home from Acapulco. This was really a shock to me, however, as I had never considered the possibility of losing a crew member during the voyage. I decided to think about the effects of this decision overnight and then discuss the alternatives the next day with Brad and Russ.

On 30 September we moved to the fuel dock at *Acapulco Club de Yate* and then to an assigned berth. Immediately after tying up, four dirty gringos went to the showers before having drinks in the patio beside the beautiful free-form pool among the goldfinches and geckos flitting in the banana and coconut trees.

The next day, bottled water was delivered to the boat, and Brad and Russ emptied them into our thirsty water tanks until eighty gallons had been consumed. Then, with remorse, we said goodbye to Irve as he left in a taxi headed for the Acapulco airport. It was time to arrange for our *zarpe* to leave Acapulco.

I thought we could leave on Tuesday, but apparently there were problems. When we had officially cleared entrance to the harbor, we had paid the migration and sanitation inspectors, but we had not cleared customs nor paid the capitán del puerto. Therefore, we were required to return the next day to see the port commissioner. Now that I look back on this situation, I realize that it was entirely my fault for "assuming" that all official matters had been attended to. Our leave papers came through on 2 October, so we could terminate our eight-day stay in Acapulco. It is strange how one's feelings change. On the day of our arrival we all thought we would never want to leave, but now the challenges of the sea were calling, and we were ready.

At 1415 we cleared the entrance to Boca Grande and set the course for 125°. How we missed Irve! It was exhilarating, though, to get to sea again. We were ready to leave the overcrowded city, with its poverty and high prices. Luckily, as we passed the Acapulco airport, we missed a huge, towering cumulonimbus cloud with its wind and rain.

When I served chicken and dumplings for dinner that evening, Brad was sick. We all had stomach cramps that day, but Brad's temperature was 101°. He tried to sleep about 1800, so Russ and I divided the time until our regularly scheduled night watched began at 2100. About 2315, while I was on watch, Russ came on deck quickly and lost his dinner over the side. Then, at midnight, Brad came up and said he felt like getting some fresh air and would trade watches with Russ. It had been raining, but was clearing, so Brad's watch and Russ's were both uneventful. By the time my turn came up at 0600, I was dry, warm, and felt much better. The sun came out to dry the cockpit and warm up our spirits. The clear blue sky with scattered cumulus clouds marked our 100-mile point from Acapulco that afternoon.

We had been standing three-hour watches and really sleeping soundly on our off-watch hours. The stomach upsets which we all suffered must have taken more out of us than we realized, because we were tired.

The winds were variable and generally light so we didn't make very much speed. The days and nights became overcast and obscured the navigational stars so we could only navigate by D.R. (dead reckoning. The positions estimated on course and speed and drift.) On 5 October, however, the wind and seas picked up so we were able to sail again and make some headway. We were visited by a red-tailed hawk that liked to perch on the baggywrinkle of the mizzen shrouds. Within an hour there were two more hawks aboard. Certainly, they must have been blown out to sea by Hurricane Fifi. The ocean became fouled with all kinds of flotsam. There were coconuts, twigs, large branches, and even entire trees. We gathered many coconuts to eat and to use in cooking. At one time I started to gather in a beautiful piece of driftwood when Brad yelled, "Wait, there is a snake." The olive-green sea snake lay coiled around one of the branches. Since the sea snakes are poisonous and antivenim is not available to counteract their bites, the driftwood was allowed to freely continue its journey, carrying its passenger to an unknown destination.

We tried to fish in this area, but were discouraged by several large sharks that kept circling the boat and dodging the debris in the water. Many birds were feeding, and several birds landed on *Pagurus II.* Many were swallows and finches that are really "land birds." While the birds were sitting on the boat, the hawks seemed to ignore them, but when they left, they were pursued by the predators. We watched one of the hawks chase a small seabird (phalerope). Although the hawk made several strikes, the phalerope outmaneuvered and finally landed on the water. The hawk made several passes about ten feet over the small bird, making no attempt to catch it, and then flew to its perch on the main spreaders. After about ten minutes of maintaining its balance on this slippery, moving perch, the hawk flew off into the distance. He soon returned with a yellow warbler in his talons. Landing on the poop deck, only three feet away from where we were sitting, the hawk leisurely pulled apart and ate his prey. Our cameras clicked rapidly as we recorded this unusual phenomenon so far from land. Our D.R. was latitude 14′ 35.0′N and longitude 98° 15.5′W.

It was also a source of excitement to spot many sea turtles among the debris in the ocean. Although we talked about how good it would be to have turtle steaks for dinner, none of us really wanted to capture and kill one of these wonderful animals.

Because of the day's heat, we often opened the forward hatch to provide some ventilation below. The hatch cover would be secured at evening in the event that changing weather or sea conditions should create an emergency. One morning, as Russ was preparing breakfast, he decided to go up forward and "dig out" some canned bacon from one of the forward lockers. I guess he must have been startled because he came up on deck shouting, "Our hawk is up there!" We tried to encourage him to fly out through the main saloon, galley, and up the campanionway, but he merely sat staring at us and opening his beak in a threatening manner. Finally, by opening up the forward hatch, he found the blue sky enticing enough to leave the forward stateroom. We really didn't mind giving space to this unwanted stranger, but we didn't appreciate his "leavings" all over my dive-bag.

When these land birds are blown far out to sea, they must become very tired because they are unafraid of humans. I was napping in the cockpit about noon one day when I felt something suddenly grab my toes. My sudden jump frightened the hawk when it had merely landed in the cockpit. Later that day, the two hawks that had been with us for two days suddenly left and flew in a straight line toward the NE horizon. Binoculars showed a freighter several miles away heading in an easterly direction. Perhaps our hawks found refuge in a more rapid form of transportation and found their way back to land. We never saw them again.

We were still unable to obtain sights on morning or evening stars, and our sunlines indicated that we were about fifty miles from where we wanted to be. On 9 August the seas became confused. Whitecaps and breaking waves seemed to come from every direction. The waves were not high (only about four feet), but it felt as though we were in a boiling cauldron with a great homogenizer stirring the waters into confusion. These conditions produce an unpredictable motion on the boat which makes it difficult to get around, read, cook, and even to rest. The wind shifted from the east to north and was very light. Since our movement through the water was slow, there was almost no steerage. To make matters worse, sudden uneven swells caused *Pagurus II* to hobbyhorse and intermittently splash several gallons of water over the poop deck. We were surrounded by debris and an occasional surfacing turtle. After twenty-four hours without apparent change of sea conditions, the engine was started, but with the confused seas, we could barely make four knots. The charts showed areas marked "tide rips," but they were supposed to be much further south than we were. Having never seen such an expansive ocean disturbance before, we could only guess that this was the phenomenon we were experiencing.

About midmorning the wind became westerly, increasing in strength enough to hoist sail and shut down the engine. It was, indeed, slow going, about two knots, but we were leaving these confused seas behind. However, our quiet was only a momentary solace.

We did enjoy a peaceful dinner and made ready to shoot any evening star that might appear through the broken cloud cover. I had no sooner sighted Vega at 62° 11.6′ when *Pagurus II* lurched so violently that I was nearly pitched headlong into the sea. I yelled, "Brad, keep her steady; I'll never be able to get these damn stars!" It was then that I realized that Brad was having a problem. The wind had suddenly come up, and yet it wasn't the temporary strong gust that accompanies tropical storms. He adjusted the sheets to try to keep us as steady as possible while I quickly located the Scorpion, sighted on Antares, and brought it down to the horizon. "Mark." It was only 29° 23.2′, but I quickly called out the angle to Russ, who was below marking the time and recording the sight. One more star; we need one more good sight. "There is one, I think it is Deneb at the top of the northern cross," yelled Brad. Wow! Amid spraying ocean wind and the rocking boat, I sighted Deneb and yelled, "Mark" at 1250 minutes, 27 second Greenwich time. It was 56° 17.4′. By the time I handed the sextant to Russ after shooting only three stars, we were logging nine knots.

Brad said that he would hold us steady while Russ and I worked through the tables in hopes of determining our location. It wasn't a perfect fix, but we were in the ballpark and recorded 10 October 1740 fix (13° 24.0′N and 95° 40.0′W). This was really the first decent fix we had since leaving Acapulco.

Outside, the wind was 35 knots with no signs of abating, so we furled the mainsail at 2100. A staggered watch with two men on gave us each two hours off out of every four, but sleep was impossible because of the building seas. By morning the strong gale winds created very high waves upon which we would rise as if we were coming out of the sea—only to suddenly plummet into the troughs. The entire sea looked white with the fine spray being lifted from the wave crests by the strong winds. It wasn't raining, but the usual bright light of dawn was obscured by the dark, sinister clouds above. The jib and mizzen were too much sail. We would have to reef. I had to postpone this decision until the light of day would aid us, but even now the situation would be dangerous. Just as Brad and I were heading forward, with the lines of our safety harnesses attached to the lifelines, we heard a sickening rip. The upper third of the jib had completely separated from the headstay and was flapping wildly, a tear separating it from the lower two-thirds.

Damn! We had tried, unsuccessfully, to get new jib snaps in Acapulco when I noticed they were wearing thin. When Irve left us to fly home, he promised to send more jib snaps to us in the Galápagos. It seems as though even with extreme care in planning, something inadvertently is omitted. We certainly didn't have a spare set of jib snaps! The jib was hastily stowed in the main saloon and I relieved Russ at the wheel.

The gale winds were NE, so we were heading S by SE, making six to seven knots with just the mizzen pushing us along. I was thinking about the "sailing directions" that indicated predominant NE trades in this area from June to November. Everything seemed correct, but the force of the winds certainly exceeded my expectations of the trades.

My dejected spirits were suddenly uplifted when Russ appeared at the companionway with a plate of hotcakes, Vienna sausage, and a cup of hot, black coffee. Confused thoughts and doubts waned and were replaced by practical plans of immediate importance.

We would have to carefully hand-sew the jib in the main saloon by taking turns when off watch. We would remove the jib snaps from my brand new Genoa jib. This was a magnificant sail, 600 square feet, newly built, and only used once in sea trials before we left Monterey Bay. It wasn't a difficult task to cut through the professionally sewn lashings of the snaps on the Genoa, but I knew that it would be a challenge to replace them as neatly.

By evening, we had made headway on the jib repair, but it was not yet ready to fly. Good stars were available in the evening, and using Vega, Antares, and Deneb, our location was computed. I wasn't satisfied with the fix, so planned to sight the morning stars the next day. In contrast to the warm day, the night temperature dropped so that jackets were required for night watch. Sleep came easily to us when off watch, however, as we were tired, and the warm blankets were comforting.

By noon the winds had steadied and the scattered cumulus clouds in the clear blue sky were indications that the storm had passed. The jib had been repaired and once again *Pagurus II* was cutting through the beautiful clear water at a comfortable 6.5 to 7 knots. The wind vane was holding us on a true course of 115°. Once again, we could read and relax when off watch.

"Look there," came a shout from Russ. We were suddenly passing through thousands of small pinkish jellyfish with flattened bells. They were only a little over an inch (three to five centimeters) in diameter. It was interesting to observe that, although the productivity of the ocean is very great, masses of young or larval forms appeared to "travel" in great numbers. We talked about this "prodigality of nature" and of the very few larval forms that would actually survive to become reproducing members of the population. It was exciting to discuss these premises upon which a theory of evolution was founded, as we were heading toward the Isles of Enchantment which Charles Darwin had visited in 1835.

The next morning, 12 October, was a beautiful day. The night's sailing was smooth and all watches could revel in undisturbed sleep. The wind had abated a bit, but we were still moving between five and six knots. In the cool, damp dawn we sighted Sirius, Capella, and Procyon. Our position was 120 miles west of the course line drawn from Acapulco to Cocos Island, but since we should soon find the westerlies, this posiion was still favorable (latitude 11° 40.0′N, longitude 94° 29.5′W).

I planned a festive dinner that evening to celebrate Brad's twenty-second birthday. The gentle sail provided an opportunity to bake a two-layered banana cake. This was decorated with pecan-butter frosting and topped with twenty-two candles I had purchased in Acapulco. I boned a canned, whole chicken and made a Polynesian sauce with fresh coconut milk and curry. This was served over rice.

With the steady winds and smooth seas offering a comfortable heel of the boat held on course by the wind vane, I set the table below with a tablecoth and dishes. This, in itself, gave a festive appearance and change from our usual one-bowl dinners hurriedly consumed in the cockpit during anxious moments of heavy seas and winds. After dinner we all joked about Brad's present, which I produced. It was an "Acapulco" T-shirt. We had been so anxious to leave Acapulco and get under way that these memories had been placed in the far recesses of our minds. But then, we recalled the wonderful glow of sighting that refuge after the trials of Hurricane Fifi.

As we were enjoying cake and coffee, Russ announced that his birthday had been on 5 October. I really felt bad about not having been more astute and remembering his natal day. I promised him a special dinner along the trip and made plans to rectify this unintentional oversight.

The next day the winds dropped, so we powered for nine hours over glassy blue seas. Since the engine had not been used for eleven days, this was a good opportunity to exercise the autopilot and charge the batteries. A dragonfly joined the ship's company. How strange, it seemed to us, that this insect should be 250 miles from the nearest shore of Costa Rica. It would fly over the boat and then land on a shroud or lifeline, only to repeat this procedure. Our emissary from land left us after a day and a half.

For the next two days, a heavy cloud cover prevented us from making a celestial fix, so our position had to be estimated by D.R. Recalculating our position, I found it amazing to realize that Isla del Coco is in the same longitude as Alabama, Tennessee, Indiana, and Lake Michigan.

On 14 October we were passing through towering cumulonimbus clouds that were accompanied by strong winds which would change direction and pelt us with huge drops of cold rain. During the day we could strip and enjoy the fresh-water shower, but during the night the showers weren't as pleasant. Another hawk joined us to spend the day.

After an almost sleepless night of scattered storms, I came out for the 0300 watch. Flying fish of various sizes littered the decks, and Russ told me how an occasional one would fly aboard and startle him to sudden alertness. Rigel, Capella, and Procyon gave us an excellent position fix (latitude 10° 23.0′N, longitude 93° 08.0′W), but our speed was disappointing. With good winds, we would make Cocos in a week, but variable winds could mean two weeks or more.

As dawn approached, the most beautiful sunrise I have ever seen was gradually revealed. Clouds of every dimension and shape lined the horizon for 360°. They were black, then blue-grey, then pink. Next their brilliant red became orange, and then the sun was up, erasing the beautiful colors and converting the clouds to their daytime white. In the lighter breeze of dawn, I thought of the saying: "Red in the night, sailor's delight; Red in the morning, sailor's take warning."

The day was cool and filled with scattered squalls. The soapy, freshwater showers were plentiful, and at the change of watch, the off-watch enjoyed the moment of coming below, drying in the warmth, and smelling the split-pea–onion–bacon soup which I had simmering in the galley. A west-heading freighter passed us about a quarter of a mile of our port side. Russ planned to harvest one of his mung bean "gardens," so we would have fresh salad and soup for dinner.

We had several visitors this day. A storm petrel tried to come aboard, and after an hour of being hit with the sail, it finally came to rest on top of the of the dinghy cover. Because of the rough seas and thirty-knot winds, it kept falling off. Once the wind blew the poor bird into the mizzen, and it fell next to me, only to flutter away again. I felt empathy when it finally left us. A mourning cloak butterfly also made two landing attempts before being whisked away by the winds. A swallow attempted to negotiate a landing and finally flew down the hatch. Brad and I laughed as we knew Russ was asleep. We wondered about his reaction if the bird were to land on him. The swallow soon appeared at the companionway and sat on the slide for a few minutes before returning below. A startled cry from Russ was heard below, and the bird joined us topside, sitting on the port lifeline.

A young tern flew into the cockpit, using one wing to steady itself against the grating as the boat pitched and rolled. He pecked our offering of flying fish, but refused to eat. After two hours, the tern flew away.

The seas were becoming more confused, and as the water

peaked up at the surface, the breaking whitecaps were blown by the wind, creating a fine, misty spray. A humpback whale sounded about forty meters off our stern. Estimates of its size were between thirteen and seventeen meters in length. As it repeatedly sounded, we could glimpse its long flippers, nearly a third the length of its body. Since its direction was not constant, we presumed it to be feeding in the area. I believe this interesting humpback *(Megaptera novaengliae)* was an adult, as this would be maximum size for this species.

Relatives of the humpback whale include the lesser rorqual, which may grow to about ten meters, and the gigantic blue whale which may reach lengths of more than thirty-three meters. These whales all feed by swimming through clouds of plankton with open mouths. They then close their mouths, forcing water out the sides and using the baleen plates like a huge filter to catch the small planktonic organisms in the fringe of bristles. The plankton is then licked off by the tongue and swallowed. Again, we were reminded of the productivity of the ocean. Whenever we were caught "in irons" with no wind, we made several vertical plankton samples. This was accomplished by lowering a net over the side, to any desired depth, and then pulling up the net slowly. Every time this was done, the net would be filled with small pink eggs as well as other planktonic forms. We believed these eggs are those of the flying fish. If we could gather them so easily in one plankton net, the humpback whale swimming with its mouth open was certainly able to gather tons of these eggs.

We watched a large dorado chasing flying fish as the boat slowly cut through the water at two knots. In the last two days we had only made eighty-two miles. I was now beginning to wonder if we should really attempt to reach the Cocos at all. We were anxious to get to the Galápagos and begin our research. These thoughts were interrupted by the arrival of a goldfinch. It seemed very "chipper," hopping all about investigating the clothespins and clothing we had hung on the lifeline to dry. After an hour and a half, its thorough survey completed, the little bird departed.

We were tiring of canned food, so we decided to try our luck at fishing. I noticed that the dorados, when chasing flying fish, would often follow their prey three feet or more out of the water. Brad attached a leader to a hook and threaded it through a small bait fish. I put this out and waited. In a half hour, the line grew taut and the reel buzzed. Brad and Russ came up on deck just in time to see a four-foot male dolphin fish jump out of the water. Suddenly each of us could taste those mahi-mahi steaks for dinner! I took the pole around the shouds to the port side and then up to the bow as I followed him. Even though I was reeling in, the star-drag maintained tension, and the fish was still taking the line out. Several times he again jumped clear out of the water. Russ had a club and Brad was ready with the gaff hook. I played him for a half hour, and he was tiring; he came to the boat and Brad was ready; then he dived. Another time, I got him to the boat, and Brad was ready, when the fish gave a sudden lurch, and the line parted. What a disappointing feeling! Inspection of the line revealed that the knot in the monofilament line had merely come undone. Brad couldn't understand it, because he had tied the knot "just like grandpa had shown him." Well, we tried again, in vain, as Russ began preparing curried shrimp, rice, and biscuits for supper.

During the night I saw many falling stars. Although most lasted only a second, two lasted more than five seconds and looked like a Roman candle with the "star" leading a long trail. As I enjoyed these spectacles in the eastern sky, I wondered if anyone else in the world was seeing and appreciating the beauty.

With little to do, our time was spent playing chess, backgammon, cooking, washing, straightening things out below (although to an observer, this might not have been apparent), sleeping, and reading paperbacks. I believe we had been reading a book a day, some of which were very good and some of which were later examined with curiosity by the denizens of the deep!

"What is that?" shouted Brad. A large object was floating in the water, so we ghosted over towards it. A large turtle was exposing its back to the sun. Turtle steaks are good, and would be a welcome change, but after a short discussion, none of us wanted to take part in the bloody slaughter which would be required. We slowly passed the turtle as the wind died completely and *Pagurus II* floated on a still, barely undulating, blue sea.

The boring doldrums afforded an opportunity to row around, take pictures, read, repair gear, and clean the bottom.

"Well, let's get some exercise." We launched the dinghy and took turns leaving one aboard; we rowed around the boat to inspect the various fittings. The barnacle growth on the hull, which had been cleaned in Acapulco, was unbelievable. We washed the topsides and then scraped the barnacles by diving under the boat. Our activities attracted several fish that appeared to partake in a free meal.

Next, we tried to catch a fish, but only succeeded in catching a small shark. It seemed as thought we no sooner brought the dinghy aboard and secured it that the wind began. What providential timing! Perhaps we have reached the westerlies, I thought. Although the wind is now from the south, maybe after this night it will swing around to the west, I said to myself. Most all the wind since leaving San Diego had been NE or E.

Our little goldfinch came back aboard in the afternoon, poked around in various corners, refused the alfalfa seeds on the foredeck, and then flew off. At dusk, a hawk made several attempts to come aboard. He once alighted on the radar reflector, but slid off. Finally, he successfully clutched the baggywrinkle on the mizzen shrouds, where he remained for the night.

The rains and increased wind began at 0245 and continued until noon the next day. A swallow that had joined us the previous night spent the dark hours perched inside the strut of the wind vane. Dawn brought more visitors. A blackish-brown moth temporarily landed on the bridge deck and was blown off again before we could capture it. A yellow canary settled on the companionway hatch under the dodger.

In the afternoon about 1530 another swallow came aboard, followed by a large hawk. The hawk made several passes, but couldn't seem to time the roll of the boat to its landing attempts. Several times the swallow would fly off, only to be chased by the hawk before returning. We watched the drama, oblivious of the pouring rain. At dusk, the swallow successfully gained its perch inside the wind vane while the hawk kept landing on the mizzen, only to lose its perch as the boat rolled. When darkness descended, the hawk disappeared.

During the night the winds picked up to a steady thirty-five knots, and we were "flying" at nine knots. This seemed to be an extensive storm system with heavy seas. After four days and five nights of continuous storm, without a celestial fix, our D.R. indicated that we were in the vicinity of Isla del Coco. However, with no indication of a change in the weather, we might have had to wait for several days. It was then that I decided to change course for the Galápagos, 420 miles away. This proved to be a good decision because we weren't to get a positive position fix for four more days and one more emergency situation.

One and a half days after the course change to 200°, the outhaul on the jib parted and the jib tore in a line parallel to the foot and about ten centimeters above the slides. This was a relatively new jib, and I now knew it was too lightly built, and without enough snaps on the forestay. I really should have had a smaller, heavier jib for this weather. This was certainly a mistake in planning.

During daylight hours for the next two days, Brad and Russ stood double watches as I slowly but steadily hand-sewed the recut jib. We were wet, cold, and tired when the jib was bent on and hoisted at 1430. The improperly fitting, recut jib did the trick, and once again we were on course for the Galápagos, the wind vane holding the course. Russ's "one bowl" dinner of meatballs, tomato sauce, string beans, and rice was enjoyed with gusto before settling down to nighttime routine.

The next day the sun occasionally shot rays through breaks in the clouds, and on the evening of 26 October we were able to get the first position fix in over a week (latitude 03° 39.5′N, longitude 86° 40.0′W). During the night we were accompanied by schools of porpoises, as if to herald good weather.

When I awoke on 28 October, Brad and Russ were busy sewing the slides onto the grommets of the main where they had become weakened. The cloud cover was still 90 percent of the sky, but spirits were rising. I prepared grapefruit juice, bacon, biscuits, and coffee for breakfast.

We were about 225 miles from San Cristobal, but we would likely be a little north because of the winds. Stowed below was a bottle of wine which Dave Hootan gave us in

Brad at the helm.

San Diego. The occasion was to be when we crossed the equator. Already, I was making plans to date it and sign it; I announced to Brad and Russ that whoever sights land first will be able to maintain the bottle as a keepsake (sans wine, of course).

Our fix on 29 October placed us about fifty-five miles east of the D.R. (latitude 02° 28.0'N, longitude 89° 52.0'W). Although our course for Wreck Bay was 175°, we couldn't make more than 135°. "But the westward current should help us," I thought. In light winds our speed now was only two knots. This day, however, became special when Brad hooked a big tuna (albacore). Tuna steaks cooked in lemon butter and served with rice was a welcome change in diet. Applesauce with freshly cooked oatmeal cookies and hot coffee rounded out the fare.

The next morning was so clear and free of clouds that I sighted on every star I knew. Then, we used Sirius, Rigel, and Capella to give us a perfect fix (latitude 02° 28.0'N, longitude 89° 52.0'W). A group of red-footed boobies were observed fishing nearby. Magnificant frigate birds were wheeling over the water and flying so close to the waves it appeared as though their wingtips would touch. They looked like giant black primitive birds, as they soared above and around us in their graceful flight. Two frigate birds spent an hour trying to land on the mast tops. They would circle, fly towards the mast, miss their footing as the boat heeled, and then try again. It was truly a "Halloween spectacle."

During my morning watch on 1 November, my salivary juices began to flow because of the aroma coming out of the galley. Brad was baking bread. There is nothing like nearing your landfall on a sunny day at sea with everything going smoothly. And then, to have a crewman bring up fresh yeast rolls with butter and jam! A warm glow of peace and contentment settled over us. The reverie was shattered at 1350 when Brad shouted, "Land Ho." I couldn't see a thing, and neither could Russ. Brad handed me the binoculars and I had a thrilling sensation, as I saw a slight mound breaking the level horizon. Tower (or Genovesa) Island is composed of basaltic lava flows and is rather flat, only extending 210 feet (64 meters) above the sea. This was the mound we saw.

At dawn on 2 November, we were greeted by hundreds of porpoises and several small pilot whales playing around the bow of the boat. Below, in the saloon, we could hear their squeaks. From above, their antics offered a welcome, as we could see Tower (Genovesa), James (San Salvador), Indefatigable (Santa Cruz), part of Albemarle (Isabela), Bindloe (Marchena), and Chatham (San Cristobal). When I was here with a group from U.C. Berkeley in 1970, I had dreamed of sailing *Pagurus II* to these islands, and now we were here. But we had to attend to business first by checking in with the authorities at Baquerizo Moreno (Wreck Bay, San Cristobal), so course was altered.

We crossed the equator at 89° 39.5′W at 0945. The "Pollywogs" were initiated properly, and Russ wrote on the wine bottle:

> It is true that *Pagurus* goes
> wherever the wind does blow.
> But whenever the occasion
> to sight the location,
> you can depend on Tiger Towle.

Apparently Brad had told him that I used to call him "tiger" when he was small.

Due to the adverse winds and the irregular currents around the islands, we did not approach the anchorage until 5 November. From Roca Kicker and Rocos Dalrymple, we got a good bearing and were very excited. Brad cut Russ's hair, and we all bathed and shaved. We cleaned everything below and wondered what it would be like just to lie peacefully at anchor, sleeping soundly, and not worrying about night watches. Hundreds of large porpoises, often jumping clear out of the water, escorted us into Wreck Bay. At 1655 the anchor was down, and at 1700 the engine was shut off. We had used 38.5 engine hours since Acapulco.

Brad made whiskey sours to celebrate, and Russ prepared the evening meal. We had shrimp creole and rice, served with sparkling burgundy.

A grim reminder on the beach in front of the Police Station informs us that we are, indeed, at Wreck Bay, San Cristobal, Galápagos.

THE GALÁPAGOS ARCHIPELAGO

CHAPTER V

GALÁPAGOS ARCHIPELAGO

After thirty-one days at sea, the quiet bobbing at anchor was almost terrifying. Had we not attempted to call at Isla del Coco, our time would have been much better. Just as dinner had been completed, we heard a thumping on the hull. An officer from the Ecuadorian navy and an aide requested permission to come aboard. Fernando Toledo was *El Capitán del Puerto, Armada del Ecuador.* He was very friendly and said that they were worried because I had not arrived at least a month before. He had been informed, on 28 June, by the *Ministerio de Agricultura y Ganaderia* in Quito, Ecuador, that we would be arriving to gather data needed for the preservation of the marine communities of the Galápagos.

After signing the official papers of entry, Fernando agreed to sell us forty gallons of diesel and informed us about checking in with *immigración* and the *policia.* He said that we were now free to go ashore, and if we had any problems or questions he would be glad to help us. We bid him *"Buenas noches,"* and then slept better than we had for a month.

The light of dawn revealed Baquerizo Moreno as a small village above the sandy beach. Behind the wooden buildings, gradual volcanic slopes were covered with small, shrublike trees called *Palo Santo (Bursera graveolens).* The pale trunks with white, twisted branches gave a ghostlike appearance to the slopes. The low, overcast clouds obscured the highlands and the little village of Progresso, which I knew to be there.

Four fishing boats were also bobbing in position, held by their anchor chains. Five more buoys were marked by small dories, waiting return of their masters.

Closer inspection of the town buildings showed the first-story construction to be of concrete bricks and the upper stories to be of badly weathered wood, with some visible indication that they had once been either whitewashed or painted. In front of the Policía we could see the old decaying remains of a forty-five-foot vessel, her hull partially buried in the sand. This was a fitting reminder of the name for Baquerizo Moreno, "Wreck Bay."

To the right of the town we could see the tall steeple of the church and the mercado, where the villagers come to buy produce and to talk. Much of the fruit and vegatables come from Progresso, where the rainfall allows a lush growth to flourish. Coffee, bananas, avocados, limes, oranges, mangoes, and papaya grow well here. Cattle occasionally are brought here from the mainland on an old rusty boat named the *Calicuchima*. Then, once a week, one or two are led down to a house by the beach where the slaughtering is done and where people gather to purchase fresh meat.

To the right, at the far end of town, we could see the *base naval* and an obelisk-type monument atop which rests a bust of Charles Darwin. This commemorates Charles Darwin's visit to San Cristobal (Chatham) on 17 September 1835. There was a landing dock in front of the pylon, and a short walk leading to the office of the *capitán del puerto*. Now we knew where we would have to go to replenish our diesel supply.

We decided to try out our "shore legs." It was soon obvious that one can feel queasy when first walking on terra firma after a long passage. It was warm, and I had many things to show Brad and Russ. We shopped at several small markets and bought eggs, papaya, avocados, and oranges; the bakery had one loaf of bread. We ordered four more for the next day. We asked several people where we could get fresh lobsters, and finally we were referred to Señor Pedro Garcia Herrans, so we decided to find him the next day.

After a breakfast of bacon, toast, and jelly, scrambled fresh eggs, and coffee, we were ready to "tackle the town" in earnest. We found Pedro with no difficulty, and he was eager to help us. I had many photographs of people, which I had taken in 1970. He helped me locate the individuals, and they were very happy and "honored" to receive these prints. I wished many times that I had known Spanish so I could have conversed more fluently, but in spite of the language barrier, we found the people in Wreck Bay to be friendly and outgoing; we felt at home. We met Pedro's son at his daughter-in-law's home. He promised to dive for some lobsters.

Pedro seemed to enjoy showing us around and even accompanied us to the church, where I wanted to introduce Brad and Russ to Brother Raphael. He maintains a natural history museum and has an extensive collection of specimens from all the islands of the Galápagos. It is a good collection, but the specimens are not well preserved or labeled. Nonetheless, this represents a monumental task and has required much effort on his part.

We then visited the *escuela* where Pedro's sister teaches a second-grade class. It was fun talking to the twelve pupils in class. We checked their math, as they were practicing additions. Then we left Pedro and walked up the hill about four blocks to the bakery. It was interesting to see the large brick kiln-type oven in which the baker built a wood fire in the early morning hours. When the temperature is suitable, the unburned wood and ash are removed and the bread put in to bake. We ordered more bread for the next day and bought some rolls for supper. On the way back through the town we met Pedro again. He took us into every store, introducing us to all the proprietors. I found Umberto, who was mixing cement for paving sidewalks. He is one whose picture I had taken previously in Progresso. His delight was obvious.

Then we took Pedro to the "best restaurant in town," which we had asked him to select. He said that we were to have a typical Galápagos lunch—not Ecuadorian! The first dish was composed of chunks of cold lobster, onion, and lemon. Then, a large bowl of fish chowder. I believe this was made with the dried, salted codfish which abounds here. It had carrots, potatoes, and beans. Next, we had a large bowl of rice with a slice of cheese on top. The cheese is bought, wrapped in ti leaves and preserved in a brine, but it tastes much like the Greek goat cheese, feta. *Naranja jugo* (orange juice) was served as a beverage. We bid Pedro *"adios"* after this very adequate lunch and rowed back to *Pagurus II.*

That afternoon Pedro's son came by with five beautiful lobsters. In fact, they were so large, I could cook them only one at a time. I set our largest kettle to boil while Brad and Russ took some five-gallon "jerry cans" to replenish the diesel. They later told me that all of it had to be filtered because it was rusty as they poured it out of fifty-gallon drums at the *base naval.* Fernando informed us that when, in our itinerary, we reached Baltra, we would also be able to obtain fuel and water. "Now," I thought, "we won't have to worry about using fuel and water during our stay in the Galápagos, because we will be able to depart with full fuel and water tanks." Here at Baquerizo Moreno, Pedro Salinos was in charge of water. He agreed to turn on the water supply to the dock at 1000 hours the next morning. He laughed when I told him I would like about fifty gallons. "Oh, about five *sucres."* Then, he laughed again. In terms of U.S. money, this would be twenty cents.

Our feast that evening was served on a cloth-covered table set with candles and flowers. Mai-tais preceded the fare and were followed by lobster, served with butter and fresh lemons, kidney bean and garbanzo salad, and bread. Coffee and fresh papayas completed this quiet repast. The few dim lights along the shore contrasted sharply with the last anchorage at Acapulco, where city lights and sounds of music and people surrounded us.

After a game of cards, contentment and peaceful sleep settled over *Pagurus II* as a fine mist decended, making the dimming lights of the village appear to fade in the distance. Baquerizo Moreno and its harbor were in slumber.

The next morning found us eager to accomplish the chores of the day in preparation for leaving. When Brad and Russ walked up the hill to the bakery, I started for *base naval* to obtain our leave papers. On the way, a terrific stomachache and nausea caused me to wonder if I could make it. "It must have been the rich food to which our systems were unaccustomed," I thought. At the office, an aide informed me to return *"mas tarde—una ora mas tarde." "Si, gracias,"* I replied, and felt dizzy and pale while negotiating the steps towards fresh air of the beach.

Still feeling uncertain about my condition, I walked to Pedro's house, as I had promised to take pictures of his family and mail them when we returned home.

My condition must have been more apparent than I had realized, because Pedro immediately sent someone to the store for Alka-Seltzer! He insisted that I drink a glass of water with two tablets, streaming bubbles and filling the water with effervescence. He also "knew" that my problem was caused by eating too many avocados!

When the photography session was completed, I met Brad and Russ, who agreed to take my camera equipment back to *Pagurus II* while I returned to *base naval.* How I wished they could have taken care of this detail for me—but "The captain must sign the leave papers." Arriving at the *base naval,* muster was in progress. Nine enlisted men (in varied attire) formed two rows, and a chief was talking to them while Fernando Toledo was standing on a step supervising the procedure. As I stood on a step behind Fernando, a sudden wave of nausea overcame me. It felt as though all the effervescent bubbles were rising from my stomach to my throat. In this precarious positoin I had visions of a most embarrassing spectacle about to occur. With great swallowing gulps, I tried to "meander casually" down the steps, to the side of the muster group, and down to the beach. I found some sand and rocks where, out of sight of the Galápagos contingent of the Ecuadorian Navy, I bent over and was sick.

At the conclusion of the muster I returned to the yeoman, who was preparing my leave papers. Then I was ushered into Fernando's inner office, where he was much more businesslike than I had ever seen him before. Several times the yeoman was summoned to retype something correctly before I signed four copies and paid twenty-three U.S. dollars.

"Now," Fernando explained, "you are provided with some unusual papers. You are free to go anywhere in the Galápagos Archipelago, and the only place where there is a port captain and entry fees are required is in Puerto Ayora on Santa Cruz (Indefatigable)." (This later proved not to be true, for I was charged port entry fees at Baltra when we entered to refuel and replenish our water. But perhaps this was because we remained there overnight.) Fernando also informed me, since we were leaving on Saturday, the entry fees at Puerto

Ayora would be more. "Perhaps you would prefer to anchor at Santa Fe [Barrington] until Monday morning to avoid extra fees."

On Saturday 9 November, at 0355, the engine was started, the anchor weighed, and we headed towards Dalrymple Rock from where a course of 271° would carry us toward Sante Fe and then Santa Cruz. We would power this short distance, as I was anxious to reach the Darwin Station after such a long delay in travel. The autopilot was set; Brad and Russ would share the watch; I went to bed below to allow sleep to complete its cure on my still queasy stomach.

At 1130 the hook was dropped in Puerto Ayora, and the stern anchor rowed out to give us a secure position. Seven sailboats and one power boat were anchored here; all but one were Ecuadorian. The black-hulled schooner *Bagheera,* from Charlevoix, Michigan, was flying a huge American flag. She was waiting for another boat, the *Beyond,* which was on her way from Panama. Then, the two would travel together to Tahiti. Our nearest neighbor was the *Beagle III,* a research vessel used by Darwin Station. We had arrived, and the excitement of accomplishing a longstanding dream caused a quickened pulse.

At noon the port captain, Señor Ochoa, paid us a visit. Paul served as his interpreter. Paul was taking part in an international program designed to acquaint high-school students with field research. He was from Kansas, and his research involved investigating distribution of terrestrial snails with Dr. Guy Cuppois from Belgium. What a marvelous experience for a budding scientist, I thought. After Señor Ochoa advised us to see him when we left port, he took his leave.

"Hi—name's Bud Divine," the voice said as an outboard drew alongside. He came aboard and accepted a glass of straight whiskey as he told us who he was. He had arrived in the Galápagos from the United States about seven years ago, and had aquired 900 acres and was raising beef cattle. We were invited up to his ranch, but this would have to wait until he returned. He was going to leave on Monday for Guayaquil, Ecuador, when the *Iguana* would call to deliver supplies to Puerto Ayora.

No sooner had Bud departed when another outboard approached and someone shouted, "Dr. Albert Towle—*Pagurus*—We were expecting you about September fifteenth." This is how we met Ralph Sievers, caretaker of "Estacion Biologica Charles Darwin." We were informed that Dr. Tjette de Vries was in the lab and would anxiously await our arrival, as Dr. Craig MacFarland, the director, was in Guayaquil for two weeks. There was also mail waiting for us at the lab! Now, among the things that will speed a sailor's visit to shore, mail from home is one.

A half-mile row brought us to the marked but narrow channel which we successfully negotiated in spite of the broaching swells. Small marine iguanas, Sally Lightfoot crabs, and a great blue heron greeted us at the landing as we disembarked. A short tour of the facilities and the invitation to use specific laboratory space really made us feel welcome.

The return trip to *Pagurus II* in the afternoon winds and choppy waters of the bay was eased by Dr. De Vries's offer to tow us with his outboard. Russ was obviously (and understandably) disappointed when he didn't receive any mail. We all knew that letters must have been posted, but all kinds of strange, unexplainable things do happen with international mail. We read our letters out loud while Russ prepared dinner. Brad, however, did not share those from Lynn!

Sunday was designated as a leisurely clean-up period. The sides, bottom, and decks were scrubbed, sail covers put on, engine oil and filter replaced, battery water added, teak oil applied to the rails, and the galley and saloon put in order. The stereo tapes enhanced our enjoyment of the work below, as a steady, fine mist settled out from the gray overcast.

We had arrived during the *garúa.* From May to December, the cooler temperatures and overcast form a mist or fine drizzle which may continue for many days before a break would admit the sun to heat up the volcanic slopes of the islands for a few days. Then the overcast and mist would return. While sitting below in the snug saloon, it seemed incongruous to realize this was considered the "dry season." The hillsides, however, were dry and brown in appearance, and the highlands were obscured by the lower layer of clouds.

Two major ocean currents influence the climate of the Galápagos. The Peru coastal current (Humboldt) flows northward from the cold southern waters, along the coast of South America. When it converges with the Equatorial countercurrent flowing eastward, it turns to the west, passing around the Galápagos. North of the equator, the El Niño current flowing south from the Gulf of Panama brings warmer tropical water to meet the Equatorial countercurrent, causing it to turn westward, usually passing north of the Galápagos Archipelago.

The point of convergence of these three currents shifts to the south during January through April because of changing wind direction. Thus, the weather of the Galápagos is affected, causing the "wet season," with a replacement of the stratus clouds by towering cumulus, bringing torrential rains, variable winds, and warmer temperatures.

Now, in November, I could picture the great convergence to be in its northward position, held there by the strong and steady southeast trade winds. I also knew that we would experience a change in the weather of the Galápagos before we left.

I thought about the meeting of these great currents hundreds of miles away and their various beginnings even thousands of miles away, bringing predictable seasonal changes to these islands, as I dove into the water for a swim. I was soon joined by Brad and Russ, as we relaxed in the cool water and shared the feeling of a productive day.

When we were preparing to leave the next morning for the library at Darwin Research Station, my eyes swept the landscape of our anchorage. To the east, about a mile from the entrance of Academy Bay, was the tiny island, barely more than a mound of sand, Coamaño (Jensen). To the south was the protective reef in back of which was Angermeyer Point, a volcanic rocky area on which Carl and Marga Angermeyer had built their comfortable home of lava rock and mortar. A similarly constructed house covered with beautifully flowering bougainvillea was close to our anchorage. Here, Gus Angermeyer and his Ecuadorian wife, Carmen, lived. Fritz Angermeyer's home sat on a gentle slope behind.

The home of Bernard Shreyer, the captain of the *Beagle III* stood out as the two storied structure further up on the slope. Beyond the Shreyer's home lived Andre and Jacqueline de

Roy (from Belgium) with their two children, Tui and Gil. Forrest Nelson (an American) lived on the east side of the bay.

We got into the dinghy and began the row along the lava cliffs and boulders to the landing of Puerto Ayora. Tall cactus *(Jasminocereus)* were growing among the rocks of the cliff, thrusting their stems fifteen feet into the air. Scattered plants of another cactus *(Opuntia)* were also visible. Thick clumps of mangrove bordered the landing near the Nimfa ("Mermaid") bar where we landed. We tied up the dinghy and began our walk through the dusty, unpaved main street from Puerto Ayora. We walked past the post office, the church, and the *base naval* of Puerto Ayora. We passed two grocery stores and an open "farmers' market" building, which maintained almost unpredictable house of business. We walked past the "commissary" which was, perhaps, the best-stocked market in town. Ralph Sievers, the manager of Darwin Research Station, is also part-owner of this store. As the road made a sharp bend to the right, we passed the public washing area. Here, the people of the town come to do their laundry in cement tubs, protected from the hot rays of the sun by a galvanized tin roof. As we passed the cemetery, the road made a sharp turn to the left, and we could see Forrest Nelson's house on the right. Another quarter of a mile and we arrived at Darwin Research Station.

The Galápagos Archipelago is shown on charts as the Archipiélago de Colón. For more than 400 years they have been ports of call for Spanish, English, and American sailing vessels. Whaling crews would often stop to collect the giant tortoises for food. The islands were named the Galápagos after these huge, shelled reptiles. Later, in 1550, Diego de Rivadeveira named them *Las Islas Encantatas* because the irregular and strong currents caused becalmed vessels to drift through the mist, creating strange illusions. In the next century, the islands were given English names, but when claimed by Ecuador, in 1832, the names were changed to Spanish. Therefore, the islands today have two names. (Many have three!)

Since the visit of Charles Darwin in 1835, scientists have become interested in these volcanically formed islands. The plants and animals living here have adapted to a harsh environment, and although many are quite different from their South American relatives, many share similarities. Speculation as to how these differences evolved and how many of the plants and animals reached the islands in the first place offers intriguing studies for many scientists. In 1957, UNESCO and the International Union sent an envoy to the islands for the purpose of selecting a site where scientific study could be conducted. Building on the east side of Academy Bay, Santa Cruz, was begun in 1960, and the first phase completed in 1964. Since that time, new dormitories and a new tortoise-house have been built.

One important consideration of the Ecuadorian government, and of all scientists, has been to conserve the unique natural beauty of the environment, the plants, and the animals of the Galápagos. And yet, these wonders should be available for people to visit. The delicate balance in nature has become a concern for the citizens of the world, and the government of Ecuador realizes the importance of conservation of this valuable resource.

We stopped at the tortoise house, where eggs, collected from various islands, are incubated, and the newly hatched tortoises fed on specific vegetation from the island of collection. When these tortoises are considered to be able to fend for themselves, they are taken to the island from where the eggs had been collected, and released. It is hoped that the population of these giant animals will again reach a level where the population will become self-sufficient. But even these efforts may be too late. Early sailors had released goats into the Galápagos. These feral goats have grazed the vegetation up to a height which prohibits the tortoises from obtaining food. Thus, the delicate balance existing for thousands of years has been upset, and the tortoises have become endangered.

It is hoped that the balance can be reestablished by killing the goats and releasing hundreds of young tortoises each year. Only time will show if these efforts have been successful. We left the tortoise house and walked through the giant *Opuntia* cacti, thinking about the preservation of these reptilian goliaths.

While we were walking to the library, we noticed that the

Iguana was preparing to leave the bay. She is a supply ship that makes regular calls between Guayaquil, Ecuador, Baquerizo Moreno, and Puerto Ayora. She sometimes carries sightseers to other areas. At the same time, the *Floreana* was arriving. She is another ship which not only brings supplies but regularly carries passengers to view the wonders of these unusual islands. She used to be of Greek registry and named the *Lina-A,* but now she is owned by the government of Ecuador. (In the fall of 1975, the *Floreana* foundered, but all aboard reached safety.)

For the next week and a half we reviewed the available literature on the marine forms found in the islands and met many residents of Puerto Ayora, as well as several visiting researchers. Dr. Craig MacFarland, the director of Darwin Station, returned with Jerry Wellington, a research worker from San Jose, California. They had come from Guayaquil aboard the *Calicuchima,* and the mail they brought with them was most welcome.

On Thursday 14 November Jerry Wellington accompanied us on a long hike to Bahia Tortuga, on the west side of Santa Cruz (Indefatigable) Island. It was a difficult hike along cliffs and oftentimes over lava-strewn boulders of various sizes lining the shoreline. Many stops were made for the purpose of examining the marine life among the boulders at the water line. At Punta Estrada I collected several specimens of sipunculans (marine worms) which I was studying. Many of the volcanic rocks had small holes, as if formed when the lava was boiling and bubbly. These were popular homes for the worms, as well as for many other marine organisms.

We were thankful for the overcast sky, without which the sun would have been unmercifully hot on the lava slopes. Many agile Sally Lightfoot crabs, with the interesting scientific name of *Grapsus grapsus grapsus,* were seen as they jumped from rock to rock to find hiding places. Fewer numbers of their darker red relatives, *Grapsus lividus,* were observed.

When approaching the north side of Bahia Tortuga (Turtle Bay) we noticed many holes in the muddy sand at the highest tide level around the mangrove roots. We sat down to watch in silence. Soon all the holes seemed to come alive as fiddler crabs appeared at the entrances, their one large claw held as a violinist might hold a violin. As they removed the debris from their burrows, they would roll up little balls and push them away. Each burrow, therefore, had several lines radiating from it, and at the end of the line was a spherical accumulation of mud. There are two species of fiddler crabs living here. The larger one *(Uca galapagensis)* occupies the more exposed areas of the sand, while the smaller one *(Uca helleri)* inhabits protected areas with finer sand particles among the mangroves.

These same areas were occupied by another burrowing crab *(Ocypode gaudichaudi)*. When cleaning its burrow, it carries sand on its chelicera (claws) like a little bulldozer. After carrying such a load about ten centimeters from its burrow, the sand is dropped and scattered as the animal "dances" on the tips of its legs. Then, the crab scurries back to the hole, and if undisturbed, only its beautiful colored antennae are visible. The light gray color of these animals and the speed with which they run over the sand on three pairs of legs, holding the last pair out behind them, and then disappear in this homes, is responsible for their name: the ghost crab.

After lunch and a swim on a long, sandy beach, we hiked along the shoreline for a couple of miles before deciding to hike overland on the return to Angermeyer Point. Although this route was shorter in distance, and perhaps easier on our feet, it was slow going. We crossed many deep fissures in the basalt and could look down a hundred feet, seeing water at the bottom. The forces which at one time separated these lava rocks, must have, indeed, been cataclysmic. Our terrestrial navigation through the Palo Santo trees, *Opuntia* and *Cereus* cacti was successful, as we recognized our location when approaching the south side of Devine's Bay. Making our way around the bay, we crossed a flat area which had been formed with many saltwater evaporating ponds. Here, the residents of the west side of Academy Bay collected the salt for use in preserving fish for the rainy season.

In our comfortable communities, we forget that the pioneering spirit still exists in many parts of the world. The migrants to the Galápagos have learned to be self-sufficient in a harsh, inhospitable environment. Without the support of supplies brought in regularly by ship, it is doubtful whether the land could support the 6000 people living mainly on San Cristobal and Santa Cruz. Even so, the residents depend heavily on their own resources and cannot afford the luxury of many items we take for granted as "necessary."

The telephone is one of these materialistic necessities. I realized this when Craig MacFarland came into the laboratory the next day and said, "Al, I just talked to your daughter." "Which one?" I asked. I was surprised when he said he didn't know. It seemed that a courier came out to the Darwin Station to tell him that someone was calling on the radio telephone. The building housing the radio and telegraph is about four miles away, and communication to the United States is relayed via satellite. I discovered, much later, that my oldest daughter, Linda, had made the call because our letters from Galápagos had not yet reached home, and the family wanted to know if we had arrived. Craig had reassured them that all were safe and well.

Excited by the prospects of communicating with home, Russ made a successful call home that evening at 1830. The cost was thirty dollars for seven minutes, but Russ felt good. Now he knew that many letters had been written and, hopefully, would arrive; everyone was well at his home in Burlingame.

The next few days were spent in working in the lab and laying in a few fresh supplies for our departure on Thurdsay for Bartolomé, a small island off the eastern shore of San Salvador (James, or Santiago). I had planned to anchor here for a day before leaving for Genovesa (Tower) where we would spend several weeks studying the intertidal marine life.

One morning we did go out to Coamaño with Jerry Wellington for snorkeling. The parrot fish, tangs, butterfly fish, and cleaning wrasses added colorful beauty to the underwater investigations. Our return was enjoyable, as we were accompanied by eight large manta rays. One had a ten-foot span, and we could see its white palps from the surface. Hurriedly, we donned our masks and fins so we could swim close to these spectacular giants. Although they feed on planktonic forms of life, they were impressive as we watched them gracefully swim by with their undulating, winglike fins. When a twelve-foot hammerhead shark came into the area, we con-

Al recording in his journal at James Bay, Santiago Island.

cluded our swim, content to observe this animal from the safety of the boat. Although I haven't read any documented accounts of the hammerhead sharks attacking humans, its squared head lobes with eyes at the margins present a frightening spectacle to a swimmer.

Though we were going to return to Puerto Ayora, we had to pay twenty dollars for leave papers and were expected to pay another port entry fee when we returned. The anticipation of new anchorages made it easy to rise at 0500. *Beagle III* left for Sante Fe (Barrington) at 0530, and we weighed anchor at 0600. Soon we cleared the reef and were heading on a northeast course. We passed Bahia Tortuga in thirty feet of water. Then we passed Nameless Island and could see Pinzon (Duncan) in the distance. Now, a course of 000° would take us to the eastern side of San Salvador.

The channel between San Salvador and Bartolomé is a narrow and spectacular one. We could see several treeless volcanic cones of Santiago. Surrounding these craters were extensive lava fields that flowed out into the water. We were entering a channel with these lava fields on either side. Then we passed close to a sandy beach on the southeastern side of small Bartolomé. Next, the channel made a sharp turn to the right, and we saw Pinnacle Rock. Near this much-photographed spire, we dropped our anchor on a sand bottom, six fathoms below, and were greeted by a flock of Galápagos penguins. These small birds were feeding. It was interesting to watch them as their bodies were propelled by the flippers, and their outstretched feet used as steering rudders. The birds were a long way from the nesting site, which is on Fernandina (Narborough) Island.

Were it not for the small amount of vegetation on the sandy strip forming an isthmus on Bartolomé, we could be on the moon! The desolate landscape was formed of visible volcanic cones and craters, surrounded by vast sheets of black lava. Nearby another boat, the *Mistral,* from Puerto Ayora was anchored. We saluted with *"Buenas tardes,"* and I could see that the captain had two tourists aboard. He must have had a very shallow draft, as it appeared as though his anchor was almost on the beach.

As darkness descended, we ate our supper of hotdogs

and chili beans, served with a salad of tomatoes, bell peppers, avocado, and kidney beans. The anchorage was good, and the wind was dying down. Our bow faced the fast current flowing through the channel, but *Pagurus II* was at ease. Tomorrow we would explore and photograph the island.

On 23 November at 0542, we left Bartolomé for Genovesa, 50 miles away. We could see the small islands of Daphne and Daphne Chica off to the starboard. The sun came out and large cumulus clouds dotted the sky. The twenty-five-knot wind made sailing very pleasant. We watched the little storm petrels *(Oceanodroma tethys)* dancing about the surface of the waves. It appeared as though they were continually "grabbing" at the water with their feet as their beaks nodded and dipped. We crossed the Equator for the second time. At 1630 we took sight of the range markers and passed through the rim of what was once a giant volcano. We entered the crater and steered for the cliffs forming the southeast rim of the crater. At 1700, the anchor was dropped in five fathoms on a sand bottom. The anchorage was good, but the wind and currents kept us parallel to the cliffs. Since the waves came in from the starboard, there was a good deal of rolling, making the anchorage uncomfortable. However, we would have to get used to this because we would be here for some time.

The next morning we rowed about a quarter of a mile to a good sandy beach. Now, we would select a site for study. Russ had planned to do a survey of the distribution and types of marine intertidal invertebrates for his master's thesis. Brad wanted to examine the distribution of hermit crabs here in Darwin Bay. We talked of the unlimited possibilities for study in the various areas as we walked along the shore. Since the tour ships made regular calls here, there were well-definied paths. Although the height of the nesting season for various birds had long passed, there were still many parents caring for the fledglings of the latest hatchings. We even found a few male magnificent frigates *(Fregata magnificens)* displaying the beautiful large red gular pouch which the females (in a receptive mood) find so irresistible.

A few nests had young red-footed boobies *(Sula sula)* being fed by their parents. Several young blue-footed boobies

(Sula nebouxii) were also in evidence. The late arrivals of the swallow-tailed gulls *(Creagus furcatus)* seemed to outnumber the others. Raucous and protective, these beautiful birds warned us to stay away from their domain as we passed. We watched a pair as they led their spotted black and white offspring down the beach near where we quietly sat. Later the family moved back up on the rocks. We watched two magnificent frigates chasing a swallow-tailed gull and even pulling at its wings in the air. They followed the gull until it landed in the water. Then, they kept diving and pulling at its wings. Finally, the gull escaped by flying to the rocks of the cliff. We have often observed the frigates giving the same treatment to young brown boobies *(Sula leucogaster).* We saw a frigate draw blood from a juvenile brown booby by raking it across the back. The frigates and other young birds take to the air by walking to the cliffs, spreading their wings, and allowing the steady wind to lift them off the ground. A frigate in the water would be helpless, but in the air their skills allow them to take fish from other birds.

We saw a couple more interesting things. A mockingbird *(Nesomimus* species) stalked and caught a fiddler crab when it came out of its burrow. I hadn't thought about this crab being part of the diet of these birds. As we passed a large sinkhole, we watched the antics of a sea lion. It had a Sally Lightfoot crab in its mouth. It kept crunching until the crab was devoured and then caught another. In books that I had read, William Beebe stated that this crab seldom, if ever, enters the diets of sea lions. Well, we now know that they do!

The *Floreana* came in one afternoon, and shoreboats carried passengers ashore. We talked with people from Michigan, who told us that General Electric had laid off 2,000 people last week. This included both management and workers. We learned also that President Nixon had resigned, and Ford was now president of the United States. We talked to another family from Quito, Ecuador, and a couple from Germany. They enjoyed hearing about our adventures. Later in the evening the *Floreana* sailed away, and we were once again alone in Darwin Bay.

Everyday the fine mist of the *garúa* season keeps everything damp. Our canopy over the cockpit had extended our living space, but occasional sunshine would have been appreciated. Although our work kept us busy, canned food was really getting monotonous. I thought when we returned to Academy Bay there really wasn't anything available to drastically change our diet—except bread and eggs, and perhaps some potatoes and bell peppers. The tomatoes were really poor, and the corn, with a red cob, was tough.

On 1 December Russ stayed aboard to do some writing, so Brad and I went ashore for photography. We hiked inland as we had planned to go over the lava slopes to the crater on this island, but as it was getting late and the wind was picking up, I felt uneasy about the anchorage. We decided to return to *Pagurus II.* On the way back, we were scaling the cliff down to the beach when we noticed a pair of swallow-tailed gulls standing near the water that had come in during an exceptionally high tide. About six inches below the water was their single egg. We might not have noticed it except for their raucous, protective clacking and crying. Under their wary scrutiny, I removed the egg and put it about ten feet away, taking a few pieces of coral "pebbles" to make a "nest." I really thought they would ignore the help and abandon their offspring rearing attempt. The next day, however, one parent bird was setting—I wonder if the egg ever hatched!

2 December was the first all-day sunshine we experienced since leaving San Cristobal. We observed the juvenile frigate birds, of various ages, sitting on their poorly constructed, twiggy nests, holding their wings outstretched, and bouncing up and down. The older ones would flap their wings very slowly, and as the wind caught the airfoil of their wings, it would lift them up. They would clutch the nest with their ridiculously tiny feet and spring higher and higher with each bounce. Suddenly one would become airborne and begin its maiden flight towards the cliffs where rising air currents caused it to soar high into the air, joining hundreds of fellow novices in the blue sky. It was surprising to see how little these magnificent frigates move their wings when soaring up and around the bay.

A pair of swallow-tailed gulls leading a spotted chick came to the higher area of the beach and sat down, watching their offspring. The young bird "trotted" on down the beach, stopping now and then, to investigate a piece of driftwood, pecking at it briefly. It then made its way to a rock about eight inches high, where it stopped and turned around to look at its parents. Then, it spread its wings and hopped off the rock into the wind. This was repeated several times. Soon, when it climbed back onto the rock, it spread its wings and began jumping up and down on the rock; at the same time flapping its wings. On one occasion, while doing this, it lifted off the rock into the wind and actually flew about two feet. It continued on down the beach into the wind, continuing the flapping, hopping, flying movements until it reached the water's edge. Here it stopped and pecked into the sand a few times, turned, and walked up to its parents. After a brief reunion, the parents flew off, leaving the juvenile alone for ten minutes before returning. The young bird then squatted very low to the ground, extended its head and beak towards the parents, and began a very "pathetic" peeping. After many peepings and poking at the parent's bill, it was fed. I often wondered if "junior" had his first real solo flight the next day. We sometimes observed the young frigates being fed by their parents. It is surprising that the parent's throat isn't cut by the jabbing of the young beaks before the head disappears entirely into the mouths of the adults.

Mockingbirds and large-billed ground finches *(Geospiza magnirostris)* hopped all around me and even on my feet as I sat on the beach. A blue band on the right leg of one mockingbird indicated a previous contact with a scientist. I watched some mockingbirds eating the fruit of the *Opuntia* cactus, some catching ghost crabs in the tidal flat, some poking around crevices at low tide, and some eating algae from the lava rocks.

I watched the small black iguanas feeding on short-cropped light green algae (*Ulva*) at low tide. Lava gulls, mockingbirds, and a yellow-crested night heron (*Nyctanassa violacea)* were its messmates.

The swallow-tailed gulls seem to be the most actively reproducing bird at this time. Some are still laying eggs and hatching chicks. The males land on the beach all day long, pick up bits of coral, and carry them to their mates. Some land and take a piece of coral, flying off without hesitation.

Others appear to be more selective, pick up a piece of coral, hesitate, look around, and then drop it for another. They make a small "clucking" noise as they return to their mates with these presents.

Once I observed two male swallow-tailed gulls engaging in an all-day battle. They were courting the same female. Their courting behavior involved pointing their beaks down, strutting back and forth, and then pointing their beaks up, making clucking noises all the while. Then, the one nearest the female would turn and peck the nearby male. Once, the male flew up and gripped the "challenger" by the wing, pulling it off the cliff. The male that seemed in control was also downwind of the challenger. He would sometimes turn around and both would fly straight up into the air, but they would be blown fifty yards or more by the wind while they were fighting in midair.

The parade of small black marine iguanas would have made an interesting movie. About noon (and it was low tide) a regular parade began as they came, single-file, down the beach from the higher area where they assembled. They skirted the east side of the beach by the cliffs where some would enter the water and swim along the lava boulders, grazing on the algae. Others would go to within two meters of the water and then turn west to cross the entire length of the sandy beach (forty-eight meters) to a tidepool where they would swim to an area for feeding. It would be interesting to know if the same ones go to the same areas of feeding every day. We noticed one animal with a large piece missing from its tail. When we rowed back along the cliff, a large eight-foot shark was swimming along the cliff in two feet of water. Many times we have seen sharks here. Also, some of the iguanas are so small it seems as though frigate birds could catch them as they swim out through the surf. Although we watched, we never saw either of these events occur.

On 8 December, having completed our field observations, I arose at 0330 and made coffee before waking Brad and Russ at 0400. By 0435 the engine was running and the anchor was brought up. Brad took us through the pass while Russ watched the depth sounder and I took bearings in the predawn darkness. Our knotmeter read seven knots, and we were assisted by a one-knot current. Our intended destination was James Bay, sixty-five miles away.

I noticed that the dishes from last night's meal were still in the sink. "Oh well, I'll just leave them there for the crew to do after we anchor tonight," I thought. At 0800 Russ made pancakes, and after breakfast I fell asleep. When I woke, the galley was cleaned, and all the dishes washed and properly stowed. I felt guilty about having had "bad" thoughts and wondered if they had somehow been transmitted. The three of us really got along quite well considering the close quarters for such a long period of time.

We had a little rain, after which the steady garúa was just enough to keep us wet when we were in the cockpit. It was also cold enough to make our sweaters feel comfortable. We could see Marchena (Bindloe) off the starboard to the north. I would have liked to stop there, but the anchorage is not very good, and I wouldn't want to navigate these waters at night. I did plan to make calls at many places so Brad and Russ could see and compare the different kinds of life on various islands.

We passed Albany Island and Buccaneer Cove on the northwestern side of Santiago. Then, we sailed past Espumilla Beach in the stiff late afternoon breeze. We could see the *Floreana* anchored here. At 1755 we dropped the anchor in James Bay, a good quiet anchorage. Because of the overlying clouds, darkness descended quickly, and by 1822 we couldn't see anything. I wondered what decision I would have made if we had been only a half hour later with our arrival! Then, I gave quiet thanks to God.

After the rocking and rolling at Genovesa, this anchorage was so calm and peaceful it was almost like being in our home berth. We all slept more soundly than at any time during the last month, and when we woke in the morning, a new and intriguing landscape was beckoning us for exploration. I pointed out, to the north, the large black volcanic flow, almost completely filling in a valley formed by older volcanic slopes. To the south of this we could see remains of buildings behind a black volcanic sand beach. Then, to the east, near our anchorage, there was a stone building on top of a high cliff. To the south, a beach of white sand offered another landing spot behind which the framework of a large building stood as mute testimony to a thwarted effort of an early enterprise. A road system and these remnants belonged to a salt mine company once in operation here. Salt was removed from a large crater, brought down to the beach, and loaded onto ships. The operation was later deemed economically impractical and so was abandoned.

About 0900 we rowed in to the beach and hiked up the abandoned, overgrown remains of a road to a spring. Although water was available, it was scanty. Years ago basins had been carved out of lava rock at the bottom of a stream bed. Except in rainy season, water barely trickles down the gully. But it is enough to keep these sumps full if they are cared for. The old rotten covers, however, hadn't been replaced by some previous and anonymous visitor. The oversight was of benefit to the feral goats and birds. A herd of ten goats spotted our arrival and quickly disappeared over the next rise. Many Galápagos ground doves *(Nesopelia galapagoensis)* ignored our presence and continued to hop around and fly into the little sump, nibbling on the moist algae. Many finches shared the water with them. We sat down to watch the parade of visitors to the waterhole.

Suddenly all the birds left en masse. They flew to nearby Palo Santo and acacia *(Acacia insulae-iacobi)* trees. A hawk *(Buteo galapagoensis)* appeared, landed on a nearby tree, and surveyed us for twenty minutes; then it left without demonstrating its hunting skills.

We stopped at the remains of the salt-mining village and examined the old rusty boilers, water tanks, and building efforts. The houses had been constructed of bamboo and must have blown down years ago. One such building was still standing. It had a kiln which occupied the entire room. This must have been the bakery.

On returning to the beach, we snorkeled to investigate the marine life, and we were soon joined by several sea lions. While drying off in the sun we enjoyed watching the lava lizards *(Tropidurus* species) which would take small pieces of Spam out of our hands. A beautiful Galápagos yellow warbler *(Dendroica petechia aureolo)* bathed in a shallow tidepool until it was frightened by an American oystercatcher *(Haematopus palliatus)* probing the surrounding crevices. Finches, marine iguanas *(Amblyrhnchus cristatus),* sea lions *(Zalophus wollebacki),* and Sally Lightfoot crabs also provided an endless spectacle as they mingled at the shoreline. While watching a Sally Lightfoot crab nearby, it shot two streams of "water" from its mouth. The unexpected jets arched out for nearly a meter. The mockingbirds here have a song, and their trill is musical.

We rowed back to *Pagurus II* as the sun was sinking into the horizon. The *Mistral* had come in and anchored close to the beach. Also, the lobster boat, *Villamil,* was making her way toward a peaceful anchorage for the night.

After supper, our quiet reading was suddenly interrupted by a loud knocking on the hull and a girl's voice saying, "Hey—anybody here?" Rather stunned, I switched on the spreader lights and we eagerly vaulted up the companionway. A pango was alongside and in it were a man and a girl." *"Buenas noches,"* said the captain of the *Normita,* which lay peacefully at anchor only a stone's throw away. They were out of matches and wanted to know if we had any to spare. I invited them aboard for cognac and gave them a box of wooden kitchen matches. "The whole box!" she exclaimed with glee. And again I was reminded of the real "luxury" of small things we so often take for granted. The *Normita* had been chartered for ten days by eight Peace Corps individuals (assigned to Ecuador) at a cost of eighty *sucres* ($3.20) per day per person—including food. They were tiring of the captain's steady provisions of fish and very little else to go with it! Laura (the visiting girl) graduated from Callison College, U.O.P.—my alma mater! Chris, Laura's boyfriend, was from Orinda, California, and was going to have his twenty-fifth birthday tomorrow; they wanted to sit up tonight and have a party, so she wanted the matches to light candles which they had brought with them. Interestingly enough, the captain's idea of escorting tourists around the Galápagos was to find a good evening's anchorage, fish for dinner, and then retire early so he could be up and away at dawn. Therefore, he couldn't see the need for having light at night.

With the promise of a recent issue of *Time* magazine and some newspapers, we agreed to join them for a while. Our contribution to the party was a bottle of whiskey. We also surprised them by wrapping two cans of chicken and two cans of spaghetti for Chris's birthday celebration the following day.

The next day we walked to another interesting "tourist attraction" of Santiago, Galápagos. After beaching the dinghy, we walked up a steep trail scaling the cliff. A path led us along the shore for about a mile where it ended at a lava outflow extending into the ocean. We walked out on a lava bridge spanning a grotto with beautiful blue water overlying white sand. As we approached, we could see many young and adult Galápagos fur sealions *(Arctocephalus galapagoensis)* sunning themselves on the rocks. Many dove into the clear water fifteen feet below. We dove in and joined them. How curious they were! They would swim around us and then come right up to us and almost touch our noses. None were aggressive, and they didn't seem to object to our presence.

We took our time returning, and walked along the beach instead of the trail, so we could compare the life-forms here with those of other islands. We watched in amazement as a yellow-crowned night heron caught and swallowed a whole Sally Lightfoot crab. The sky was blue without a trace of clouds and the sun was hot, so when we reached a tidepool with natural "bathtubs" of smooth lava rock, we bathed, languishing in the cool water amid the extraordinary companions of birds, iguanas, small colorful fish, and a lava landscape.

The *Calicuchima* arrived, and we passed two shoreboats as we were rowing back to our floating home. "Hey, these are the Americans," we were greeted as we passed. We were now familiar with the island tour boats, and the captains of these ships seemed to enjoy telling their tourists about this small "research vessel" from Santa Cruz, California.

Our run the next day would be only six or seven hours. So far, the studies had taken us to places with which I was familiar, so I was anxious to share a new experience with Brad and Russ. On 13 December we left James Bay at 0620 and the steady twenty-knot wind provided us with the force for a smooth sail under an azure sky. At 1230 we cautiously proceeded through a narrow channel with rocks marked on the chart but invisible below the surface. We threaded our way into a bayou surrounded by mangrove trees above which extensive lava fields reached as far as we could see. This was

Bahia Cartago, located at the isthmus on the west side of Isabela (Albemarle) Island.

It was the warmest anchorage we had yet; it was hot! I hoped the gentle but steady breeze would also continue through the night. A glance at our surroundings indicated that without the breeze, we would not only be uncomfortably hot, but the mosquitoes would probably feast all night. The anchor was dropped in three fathoms of water on a sand bottom. I could watch it dig in as the scope of the anchor line was payed out and the engine reversed to hold us fast.

No sooner had the engine been turned off when large manta rays, with ten-foot spans, began jumping all around us. Jagged lava fingers projected into the water and many were bordered with mangrove trees. A large turtle *(Caretta caretta)* swam by as I stood on the deck surveying our surroundings.

We rowed to the shore and tied the dinghy with both a bow and a stern line to keep it off the sharp lava rocks. We only walked over this unbelievable landscape for about a half mile, as the going was rough. The surface was composed of varying-sized jagged boulders covered with sharp, shinglelike pieces. When a piece would slip under our feet, it sounded like glass as it slid down a ravine and shattered. We crossed deep fissures where the lava had separated. Many had salt water in the bottom and were ringed with white where water evaporation had left deposits of crystallized salt. We passed large saltwater pools formed from seepage of the ocean through channels and interconnected lava tubes. Leaving our tennis shoes on for protection against the sharp lava, we enjoyed a swim in some of the large pools. After thorough investigation, we found areas where we could climb up on a higher rock and dive into the water.

The next day we packed a picnic lunch and went for a long row through the many canals lined with mangrove. It was like being in a huge swamp with the mangrove branches touching overhead. Where the channel widened, islands of mangrove trees were in the center so it was as if we were in a maze. Rays and turtles were all around. The mangrove growth was so thick that we couldn't see the volcanic rubble that we knew was located on the other side of the trees.

Hundreds of grey fish, about a food long, swam by. In the back waters we didn't see the large manta rays that were so plentiful at the entrance, but we saw many black and white spotted stingrays with long tails. Green sea-urchins were located among the mangrove roots amid thousands of small rock oysters. We soon learned that these oysters could be shelled, and the edible portion was about the size of the small Olympia oysters of the Northwest Coast of the United States. They were good raw and cooked, but required a good deal of energy to extract from their sharp, cuplike shells. As we sat in the dinghy tied to mangrove prop roots having our lunch, the peaceful stillness was broken only by a sound as if pebbles were being bumped together under the water. Occasionally, a louder sound could be heard, as if a hammer were hitting a rock. Many stingrays and one six-foot white-tipped shark glided by as we sat discussing this sound phenomenon. Although we couldn't determine the cause of the eerie noise with certainty, we decided that it must be created by tidal fluctuations entering (or leaving) the interstices between the lava boulders all around us and beneath the water on which we were floating. I think a better name for this area would be "Bay of the Laughing Water" or "Bay of the Singing Rocks" because of the constant and mysterious noises.

About 1800 we returned to *Pagurus II* for a supper of freshly shucked oysters. These were dredged in a mixture of corn meal and flour before being fried in canned butter from New Zealand. Served with lemons and a tomato sauce, and accompanied with string beans, this was truly a gourmet affair. Dessert was a chocolate cake with mochoa icing (which Brad had made) and coffee.

That evening when we retired and the lights were turned off, we heard the annoying buzz of several mosquitoes. They seemed to prefer Brad and Russ until I suggested using insect repellent, which remedied the situation. The buzzing continued, and the loud slapping sound of the jumping manta rays were heard, but sounds soon faded as sleep arrived.

The rock reef in Bahia Cartago was one of the most interesting areas I had seen in the Galápagos. It seemed to be a haven for various kinds of echinoderms. Every small coral and lava sandy patch, at the intertidal level, was covered with small black brittle stars. They would scurry away seeking refuge every time a rock was turned over. Ten-colored pencil sea urchins and black-spined starfish were everywhere. Many apple-green sea urchins with short, black, tube feet filled the spaces between the lava rocks where they were "disguised" by the many bits of shells covering them. Small pinkish or vivid red starfish with rays of varying lengths were prevalent. Large flatworms could be seen gliding away when a rock was overturned. Many of these were black, but several were a brilliant blue with white and gold spots. Large areas were covered with powder-blue encrusting sponges, interspersed with a tannish globular sponge with sharp glassy spines. Small, rectangular-shaped clams lived in the sand between rocks. We found one as large as my fist. Occasionally, an octopus would slither away as a rock was removed to have a better look at some beautiful yellow and bright orange nudibranch. Tube worms with brightly colored, feathery gills could be seen just below the surface of the water, and scattered throughout the rocks of the intertidal zone were hundreds of small (one centimeter in diameter) bright orange sea anemones.

Inshore from the rock reef which we were investigating were the mangroves, their roots and prop roots covered with rock oysters exposed by the low morning tide. The level of the high tide could easily be determined by the height at which the leaves of the mangroves began. As we removed several clumps of oysters, many small, greenish shore crabs *(Pachygrapsus transversus)* scurried away. Many empty clam shells were filled with siltlike sand and provided shelter for several marine worms.

On 17 December we arose at 0400 and prepared to leave mystical Bahia Cartago to return to Academy Bay. We sat in the predawn darkness, sipping coffee and waiting for light to aid in negotiating the tricky and rocky channel. It felt as if we were going home, because we would again see friends and collect our mail. By 0630 we were sailing past the volcanic terrain of the southern portion of Isabela on a course of 104°. Brad and I were in the cockpit anticipating the breakfast being prepared in the galley by Russ. Although they looked and smelled a bit like dog food, the corned-beef-hash patties were consumed with relish.

At 0900 we were joined by a dozen porpoises. These were

large animals, about ten and twelve feet long. Many actually touched the bow and sides of the boat as if they were "scratching" their backs. Their squeaking voices could be heard as they frolicked alongside, several actually doing barrel-rolls as they darted into the air and away from the bow. At 1530, the anchor was dropped in Academy Bay and once again *Pagurus II* rode silently next to the *Beagle III.*

After formal entry into Puerto Ayora was completed, our mail was obtained. In a long letter, Nan explained that she had had Pete Sutter (in Sausalito) build a beautiful new jib. I was concerned about the condition of the one tired, many times repaired jib we were using and was certain it would not last for the 3,100 miles to the Marquesas. In the letter, my fears were confirmed. Several packages sent to us in the Galápagos never arrived. Fortunately, we did receive two packages of jib snaps, sail thread, and shackles. It seemed as though duty was required if the sail were to be shipped to us in the Galápagos, and the fee would be equivalent to the price of the sail. Could we be assured, even then, of its arrival, I wondered. Two days later I sent a telegram:

NAN
WILL ADVISE WHEN AND WHERE TO SEND SAIL IN TAHITI
FOR BRAD PLEASE SEND LYNN AN OLD-FASHIONED BOUQUET FOR CHRISTMAS.
LOVE AND MISS YOU
AL

The holiday season in Academy Bay, as in many places in the world, presented an aura of anticipation. Decorations were brought out and displayed. The entrance of *base naval* of Puerto Ayora had a large leafless tree decorated with tinsel and ornaments; the Nimfa Bar utilized two large tables to construct an impressive manger scene depicting the birth of Jesus; stores brought out special clothing and candles for sale; the parish house displayed a large artificial Christmas tree with colorful decorations for all to see; the tree in the town square was adorned with colorful lights.

Pagurus II had to have a reminder of the joyous season. We cut a Palo Santo branch and put it on the sideboard in the main saloon. Strung popcorn served as garlands. We made colored balls of origami paper and hung them on the branches. A few small packages mysteriously appeared, wrapped in unusually attractive papers. We thought of home, our families, and friends, and knew that they were also thinking of us.

The season was also marked by parties. Forrest Nelson had an open house and we were invited. It was well attended, the dancing and festivities continuing throughout the night. About midnight, however, people were looking for more to drink. Jack Nelson drew a fifth of clear liquid from a fifty-gallon drum. He said it was "white lightning" made from potato peelings. It smelled like 90 percent ethanol! Some guests began drinking it straight, because there was no more orange juice. I didn't try it, as I didn't want a headache the next day.

One afternoon we spent an enjoyable "coffee hour" with Marga and Karl Angermeyer. Marga's mother, who is ninety-two, was with them. Until fourteen years ago, Mrs. Bronosky lived in Chicago. When she required more care, however, she came to Academy Bay. Her forgetfulness kept Marga and the maids alert, and she is a determined woman. Once she was missed, and after a frantic search, found where she had fallen, as she attempted to walk out onto the reef. Although her hearing is not keen, she is very attuned to what is going on. One could sense her deep affection for Karl as they shouted at one another loudly in a teasing manner. He would turn to Mrs. Bronosky and jokingly shout, "Frau Müller." Then, she would say, "Ya, Herr Büller." Next, they both laughed.

Jacqueline De Roy and her daughter, Tui, made jewelry and sold it to tourists to help support the family. Their artistic talent was obvious. I bought some pins and earrings of black coral which I planned to take home for much-belated Christmas gifts. Tui is also an excellent photographer, and many of her pictures have appeared in the *Pacific Discovery,* a publication of the California Academy of Science.

Both Tui and Gil are fluent in French, English, Spanish, and German. They serve as tour guides on many of the ships around the islands, and certainly they must have friends in many parts of the world because they are so pleasant and friendly.

December 24 was an enjoyable day. We worked at the Darwin Biological Station in the morning. Then, we gathered our buckets (which doubled as shopping bags) and walked through the village, purchasing tomatoes, crackers, rum, bell peppers, chocolate bars, and some potatoes. The garúa season was just about at an end, and the hot season with scattered thunderstorms was beginning; the strong, steady winds were being replaced by long calms. This made the row back to our boat an easy one as we glided over glassy waters.

In the afternoon we had a lively volleyball game at the De Roy's. The sandy court provided good exercise when running for a ball. We felt good when we returned to our dinghy and dove into the cool water for a refreshing swim. We used shampoo to wash our hair while treading water. We discovered that we could get our hair clean this way, and then a pint of fresh water when we got out would remove the salt.

Jan and Jerry Wellington were coming out for dinner aboard *Pagurus II* on this night, so we began preparations. We had cocktails in the cockpit as the sun dropped over the cactus-silhouetted skyline. For dinner we had creamed shrimp made with a light cheese sauce containing sauteed bell peppers and onions. This was served over rice, with asparagus as a vegetable. Brandied pears and peaches were served with coffee for desert.

After dinner we relaxed in the cockpit in the warm night air, and I entertained with a harmonica and the melodica. We noticed a fishing boat approaching our anchorage towing an eighteen-foot sailboat which didn't have an engine. It was most astonishing to us to notice that the small sailboat was flying a British flag.

The next morning, 25 December, we saw that our eyes had not deceived us. Sure enough, the British flag was flying on her stern. As we approached in our dinghy, I saw that the little eighteen-footer was a British-built Caprice, and her name was *Super Shrimp.* Her color, a light pink, fit the name. I invited the couple for a Christmas brunch of bacon, eggs, and fresh biscuits and jam, and they gratefully accepted, as they were out of food.

At breakfast we listened to their interesting stories of adventure which you will, perhaps, someday be able to read in Shane's book. Shane Acton had left Cambridge, England nearly two years before. Alone, he had planned to sail the South Pacific. When he reached the Panama Canal, the *Bagheera* (which was in Academy Bay when we arrived earlier) towed him through. It was in the Canal Zone that he met Iris Derungs, a Swiss girl, and she joined him to share the adventures of *Super Shrimp.*

Off Colombia, one of the sails tore so they took the boat to a beach on a little island. The island was a Colombian penal colony. The inhabitants were delighted to receive the first visitors in more than three years! The guards and prisoners all helped repair the sails on their sewing machine.

When *Super Shrimp* continued her voyage, the wind let up just north of Guayaquil, Ecuador. They would have been carried onto a reef by the currents, but a fisherman rescued them by towing the little vessel to Guayaquil. Apparently Shane's papers from Colombia to Ecuador had been lost, but they were allowed to remain in Guayaquil for two weeks and were given "excellent treatment."

Proceeding without papers, the *Super Shrimp* was becalmed outside Academy Bay, and the currents were carrying them away. On 24 December, the Ecuadorian fisherman towed them to a safe anchorage, and here they were.

I asked Shane about his charts and he said that, since space was so limited, he didn't carry many. Also, when he was through with a chart, he would just throw it away. He had planned to go on to the Marquesas, but had no charts of these islands. With interest, he studied my charts of the Marquesas and made some penciled notes. As I rowed Iris and Shane back to *Super Shrimp,* I wondered if we would see them again in the Marquesas or in Tahiti. (On Thursday 7 October 1976, the *San Francisco Chronicle* had a picture of Iris and Shane aboard *Super Shrimp* as they sailed into Brisbane, Australia. Shane had ended a four-and-one-half-year, 15,000-mile voyage.)

We were invited to participate in a shared Christmas dinner at Darwin Biological Station, so Brad baked a lemon cake, and we brought a bottle of whiskey. The hired help were given a holiday on Christmas, so the twenty-two of us present divided the chores of cooking and cleaning up. The mélange consisted of three chickens, a little meat, lots of bread dressing made with bell peppers and tomatoes, mushroom gravy, carrots, sprouts, lettuce salad, and grapefruit juice. It was an enjoyable Galápagos and international Christmas celebration.

On 26 December the *Royal Viking Star,* a beautiful passenger ship, was standing off Coamaño Island and Señor Ochoa, all in white attire, was taken out by his aides (also in white) to perform his official duties. Later, as we walked through the town, we met many passengers walking through Puerto Ayora. It was as if the town was inundated by a population that far exceeded the residents. We later learned that the tourists numbered 450. Many local people had set up little tables and signs that displayed and advertised all kinds of curios for sale. We felt that we were observing a tour from the eyes of natives rather than those of passengers.

This was the day we had planned to hike to Bella Vista, a small village in the highlands of Santa Cruz. We left the main street of Puerto Ayora and the tourists when we entered a small footpath behind the powerhouse. This was the beginning of the back trail up the mountainside. The steepness increased as we passed through a large cactus and Palo Santo forest. Soon we noticed kalanchoe *(Kalenchoe pinnata),* a small succulentlike plant growing along the trail where moisture accumulated. The pale blue-flowered shrub *Plumbago scandens* lined the trail, and soon grass and vines were on the ground, which was damp. We were surprised at a bend in the trail to notice that a fence was installed to keep cattle in desirable pasture land. Beautiful red-flowering hibiscus were everywhere. We passed through a swampy area where the grasses were knee-high, and Spanish moss festooned the trees and shrubs. *Viburnum* with white fragrant flowers that hung in clusters were scattered among coffee bushes *(Coffea arabica).* The fragrance of these blossoms faded as we entered the *Scalasia pedunculata* zone. These scalasia trees formed a welcome shady canopy keeping the hot sun off our heads. We saw large areas of angel's trumpet *(Datura arborea)* with their large, heavily scented white flowers.

Russ and Brad on way to Bella Vista, Santa Cruz Island.

We passed the Bella Vista mission and school and then walked through the one-block main street of the village. Here we came out onto the one major road of Santa Cruz Island. This road connects Puerto Ayora with the channel across from Baltra. When tourists arrive via Tame Airlines, they take a boat across the channel to waiting trucks for the drive to Puerto Ayora. Bananas, papaya, and avocado trees were scattered along the road. We stopped for a lunch of papaya and avocados, which we had under some tall balsa trees.

About 1400 we decided it was time to start back because the trail would be impassible in the dark. We walked rapidly, and two hours, fifteen minutes later we came to Darwin Station. A half-hour row brought us back to *Pagurus II.* Brad and Russ went to the De Roy's to play volleyball, but I stayed behind for a refreshing cool swim, after which I fell soundly asleep. The next thing I knew, Brad was waking me for dinner and announced that it was 1930. I think I could have slept all night.

On 27 December, we checked out with the port captain in preparation for a visit to other islands. By 1107 we weighed anchor. Since the wind was light, we powered to an anchorage in a channel between South Plazas and North Plazas, arriving at 1405. It was hot! A swim in the cold water was reviving, as we watched two powerboats, *Cathchar* and *Laura Marie* from Puerto Ayora, arrive with ten tourists. Tui and Gil were aboard one of the boats as guides. We decided to forego a visit to the island until morning so these tourists could enjoy the isolation and freedom from persons unknown.

Small finches continually visited us here. They would flit from shroud to shroud before darting towards shore. Many swallow-tailed gulls live here also. And several red-billed tropicbirds *(Phaethon aethereus)* could always be seen overhead, their long white tail feathers streaming. The sea lions were the noisiest ones we had encountered. We watched and soon discovered the large beach master. He continually patrolled his territory, maintaining order and keeping out intruders. We watched as the tourists landed timidly on a cement platform, afraid to walk the short, narrow, causeway to land because of the sea lions. I knew that we would probably be just as cautious because a barking sea lion one foot away shows very large canine teeth and, after all, we were invading his domain.

The next morning we did go ashore. As we walked along the shore we noticed the many very young sea lions. The smallest ones looked at us with wide, baleful eyes and then began to yowl with sounds like, "Oi—oi." One large sea lion had a lava lizard *(Tropodurus)* on its back, running back and forth, catching flies. The flies were all over the poor beasts, concentrating on the eyes of the young and adults alike. We discovered that the nonterritorial bachelor bulls were occupying the eastern cliffs. Here they practiced mock battles, as they were maturing and would soon fight seriously for a territory and a harem of their own.

Also along the eastern cliffs were nests of the Galápagos storm petrels *(Oceanodroma tethys)* and blue-footed boobies.

The golden land iguanas *(Conolophus subcristatus)* on South Plazas were beginning to show their mating colors. We watched them feed on the red succulent *Sesuvium* that formed low patches on the island. Large *Opuntia* cacti dotted the terrain. Unable to resist a forbidden act I took fruit from the cactus and hand-fed some of the iguanas. We examined many iguana burrows along the shoreline above the high-tide level. At the entrance to many burrows were piles of Sally Lightfoot skeletons. Although *Sesuvium* forms the major part of the diet for these large reptiles, it appears as though the crab might form a supplement.

We rowed to North Plazas Island and climbed the rocky cliff to the top where we were met by two yellow-crested night herons and a pair of screaming swallow-tailed gulls. It was impossible to walk far on the western side because of the dense shrubbery. Climbing back down the cliff to where we left the dinghy, we examined the intertidal organisms and I collected some marine worms for study.

On 30 December, after two full days on Plazas, we left for Baltra (and Seymour). As we progressed northward along the eastern shore of Santa Cruz, we passed the *Floreana* going south, probably to Española (hood) Island. At 0955 we cautiously approached the Baltra fuel dock at Aeolean Cove. This was obviously designed for large ships. With fenders in place, we tied up to the huge beams as the gentle swells alternately lifted and lowered our boat, now dwarfed by the pier and the landing—thirty feet above our decks. It took a good deal of coordinated cooperation amid sounds in both English and Spanish to turn the various valves from impressively large pipes to start the flow which delivered fifty gallons of diesel. But now, our tanks were full, as was our twenty-gallon deck load. The cost was 375 *sucres.* We then took on 100 gallons of water through a similar ritual, and a cost of 100 *sucres.*

I felt better as I directed *Pagurus II* away from this hazardous fuel pier and dropped the anchor in two fathoms of azure water onto a good sand bottom. The anchorage was so calm that the scope of line payed out merely hung straight down after the hook was set.

We landed the dinghy on a beautiful white sandy beach amid foaming but gentle breakers and began a walk along a road built by the U.S. armed forces during World War II. We inspected many bunkers with thick iron doors. These hillside storage areas must have held ammunition at one time. They were locked, now storing supplies for the Ecuadorian navy base located here.

We were walking across the Baltra airstrip when a large truck came speeding towards us. *"Buenas dias,"* we were greeted by Marcello and Carlos (a meteorologist). They were going across the island to the channel to meet a pango carrying two people from Puerto Ayora who were going out on tomorrow's flight to Guayaquil. Would we like to go with them? Sure. We were friends even before a stop was made at the "Galápagos Service Club." Here were the remains of what was once a beautiful building with tile-floored dining rooms, bar, and offices. This was once a popular retreat for American servicemen stationed here during World War II. Most of the roofing tile and much of the flooring tile had been removed by residents for their personal use.

Continuing our ride among scattered Palo Santo trees and dry grass, we approached the cliffs of the channel. Down a steep grade filled with ruts and rocks, the truck descended to a clearing almost at the water level. Mangroves lined the channel, and we could see the Santa Cruz road ending in a similar clearing on the other side of the channel.

After nearly an hour's wait, Marcello said, *"acaso man-*

aña." With that, we climbed aboard the truck and returned to the cantina where Marcello produced a pint of whiskey and introduced us to many of his friends. Carlos explained that Marcello's father was "head of the navy in Guayaquil." Ping-Pong, pool, and records provided an afternoon's entertainment. One of the men I met was the chief of naval operations at Baltra. He had been out fishing and just returned with his catch, which he was delivering to the kitchen. His parting comment was, "Enjoy yourself—*mi casa tu casa.*" "*Gracias,*" I replied. This was the first time that expression had ever been used to me, and I thought it was most graceful as a host's expression.

Marcello and Carlos insisted on driving us back to the beach, so in the early evening we climbed into another truck that had a wooden bench for a seat. The truck was so wide that the five of us sat together on the bench of this "Tame Airlines"—labeled vehicle. Marcello left in a cloud of dust. When we reached the far side of the runway, Marcello shouted, "Take off." For the entire length of the runway, we rumbled, jiggled, and bounced along at 120 kilometers per hour!

"See you at the Ninfa Bar," they said. It seemed as though most of the young air force and navy men from Baltra were going to Puerto Ayara for an *Anño Nuevo* party. If it is anything like the Christmas eve celebration, I thought, they will drink and sing all night. Somewhat relieved from the ride back, we bid "*Adios*" to our friends as we launched our dinghy into the placid waters of Aeolean Cove.

The next day we extravagantly used some of our fuel to power through calm waters for six hours on the return to Academy Bay. The *Beagle III* was near the Ninfa Bar under the mangroves on on the sandy beach. This was the drydock.

We received two air-mail packages postmarked 12 October. One contained shackles of assorted sizes and the other, a box of cookies, complete with weevils!

The town's celebration was a huge success and was completed by the burning of effigies of the major local characters. In Puerto Seymour at Baltra, we had also noticed the effigy of the port captain, Juan Garzon A.

For the first time we noticed people aboard the *Finn.* This small Swiss-registered sailboat had been anchored here since we first arrived at Puerto Ayora. I rowed over, introduced myself, and asked them aboard *Pagurus II* for cocktails. They hesitatingly said they didn't have time, so, thinking they did not understand, I asked them again in German. "Oh, ja," they both replied. We met Brunhild and Rolf Seeler. He works in Buenos Aires and in Germany. Every year they sail the *Finn* for three months and then leave it in someone's care. They had just returned and were getting ready to sail to the Marquesas and Tahiti, where they planned to find someone to look after her for another nine months.

On 4 January, the passenger ship *Stella Maris* arrived. She had left Puerto Morelos in the Yucatán twelve days ago with 165 passengers from Chicago. We had received word that Dr. and Mrs. Maurice Gleason and their daughter, Joy, would be aboard. Although I had never met them, they are close friends of a collegue, Dr. and Mrs. Robert Thornton. Bob had written and thought it would be fun if we met them at Puerto Ayora. At 0700 we rowed to Darwin Biological Station and waited for the shoreboats which were following Craig MacFarland through the tricky channel to the landing. The Gleasons were aboard, we learned, and would be coming in with the second group in about an hour. We were waiting on the dock and greeted Elizabeth and Joy when their shoreboat docked. Apparently, Maurice had a tightening in his chest, so thought it best to stay quiet for a few days, thus he decided not to come in. Joy is a linguist and teaches at Cornell; Elizabeth has a Ph.D. in Library Science. We walked through the tortoise culture rooms and showed them around the Darwin Biological Station. Then, we visited as we walked through the village to the quay where the shoreboats met the passengers to take them to Marga and Karl Angermeyer's to see the iguanas.

At the same time of the *Stella Maris's* call, the *Iguana* was anchored in Academy Bay with German and French tourists, so the residents of Puerto Ayora were really busy displaying and selling shells, jewelry, Galápagos T-shirts, and black coral.

We left Elizabeth and Joy at the quay as their time was not sufficient to allow a visit to *Pagurus II.* After the Angermeyer visit, all the tenders circled *Pagurus II,* and we waved farewell. I'm sure all the *Stella Maris* passengers were comparing our adventurous little craft flying the American flag to their luxurious staterooms aboard the Greek-registered passenger ship. It was difficult to realize that in two day's time the people would be in Guayaquil, and in another day would be in their homes in Chicago with many purchased souvenirs of their trip.

Because of the cloud layers, the morning sun was well above the horizon before it began to show. Therefore, the mornings were usually a blaze of golden hue before it suddenly became light. The days were much warmer as the thin cloud layer burned off around 0900, creating clear blue skies. Then, around noon, large cumulus clouds began to form. During the very hot afternoons, an occasional cloud would pass overhead and a sudden, pelting rain would occur. The cloudburst stopped as suddenly as it started, leaving a hot, humid atmosphere. Thus, the rainy season of the Galápagos had begun. The back trail to Bella Vista would now become a muddy, and almost impassible, path.

We scrubbed, painted, varnished, and reorganized our supplies in preparation for departure. We wrote letters in answer to our latest communications, but still the packages of needed hardware did not arrive. While at the post office, the clerk asked if we knew a specific name for whom she did have a package. I inspected the package and replied in the negative. The name and address of this package (sent from Los Angeles two years ago) were not decipherable. Furthermore, it must have contained food, as there was a hole in one corner of the package, and the contents were entirely gone. "Now," I thought, "I know the fate of our cheese and beefsticks which Nan had sent us for Christmas!" (I might add that our own postal system seems to be just as inefficient. It has been nearly three years, and we have not been reimbursed for the packages which Nan sent, insured, and which I never received!)

My completed report on our activities was typed on a Spanish typewriter before rowing back to our boat. We passed a large sea lion actually playing with a fish. It would come to the surface, shaking the eighteen-inch catch, and then dive again. Two pelicans were swimming around it in anticipation of some scraps, and nine frigates kept diving in attempts to rob the sea lion of its prize. A school of sixteen rays swam by. As we neared the anchorage, an eight-foot hammerhead shark leaped clear out of the water and did a barrel-roll before disappearing below the surface. Blue-footed boobies were diving for their evening meals.

We met Gus Angermeyer rowing his boat in a standing position, as he often did. Gus told us that earlier in the day he was rowing out in the south channel when a big head with a staring eye came out of the water only ten feet away. Gus explained: "The killer whale greeted me saying, 'Mister King, how are you today?' And then he showed me his tail and came up again saying, 'Mister King, I am going now.' " The remainder of Gus's explanations of his fear cannot be put into print!

That evening we were invited to dinner aboard the *Betty J* by Betty and Leonard Pratt. Leonard put in a phone patch from a contact in Mill Valley, California, so we were able to talk to everyone at home, exchanging holiday greetings and explaining our travel plans. When we rowed back to *Pagurus II* about midnight, we talked about leaving and the mixed emotions felt by each of us. We had met many friends in the Galápagos. The interdependence of people on one another was refreshing and very obviously in contrast to a busy cosmopolitan area where individuals bump and shove in crowded conditions, never noticing another person.

Obtaining supplies in preparation for leaving was not an easy task, as Puerto Ayora's cupboards were bare. The town depends on supply ships from Guayaquil. One of the island supply ships, the *Presidente* had been in Guayaquil, loaded and ready to sail, for many days, but wasn't allowed to leave. Forest Nelson explained to me that another supply ship, the *Pingüino,* is owned by Miguel Castro who has a brother in the upper echelon of the Ecuadorian navy. Apparently, every time the *Presidente* is ready to sail, she is inspected by the navy and found to be "unsafe for sailing." The delay allows the *Pingüino* an advantage of time, but also causes a hardship on the islanders. Puerto Ayora was out of almost everything. Only the bakery had flour, and they limited the number of rolls sold to each person. For this reason, I had decided not to buy any rolls from the bakery. The town folk accept these hardships as "the way life is." Furthermore, when the *Pingüino* arrived, she didn't bring many supplies, as she was unloaded in only four hours instead of the customary two days. The Nimfa Bar, for example, had been closed for a week. Empty soda water and beer bottles were stacked outside waiting to be exchanged for full ones. The few drums of diesel brought in by the *Pingüino* were all brought to the construction company, so people couldn't get fuel here. Local fishermen and the small tour boats, however, could go to Baltra for diesel. A little lumber was unloaded, and we saw individuals tie several planks together and tow them shoreward behind small pangos. If a *mercado* owner complains about the cost of transport, then he is no longer brought any supplies. We watched bawling cattle winched up in a sling, swung over the side, and lowered to a barge. Later, some were lowered into the water and towed behind small boats.

On 14 January, a beautiful Mariner-35 ketch, *Honeywind II,* from Newport Beach, California, dropped anchor in Academy Bay, only twenty-five feet from *Pagurus II.* We invited Janet and Jim McNabb and their crewman, H. Blake Chatfield, for dinner that evening. We enjoyed exchanging tales of adventure. They had spent three years equipping and preparing their boat and themselves while using it for weekend charter. Jim attended night school studying diesel mechanics, while Janet attended a course in radio. She said that the first time through involved memory to pass the tests, so she enrolled a second time and began to understand the function and procedure involved in radio communication. They had been voyaging for over a year, traveling down the coast and through the Panama Canal to see the San Blas Indians. Then they returned through the canal and called at various ports in Costa Rica.

Their friend, Blake Chatfield, joined them in Panama. Blake is a travel writer and planned to meet his wife in Papeete and write an article for the *Chicago Tribune.*

16 January. 0840. We were just one hour out of Puerto Ayora, heading for Post Office Bay, Floreana (Charles) Island. Although elated to be under way again, the fading of Santa

Cruz (Indefatigable) Island in the distance brought some sadness. We were leaving friends. Even the boats of Puerto Ayora: *Mistral, Normita II, Cathchar, Lora Marie, Encantata,* Fritz's *Nixe,* and Karl's *Cymbol,* were old friends we had seen at many Galápagos anchorages. *Pagurus II* was also saying farewell as she met the gentle swells, carrying us to new adventures in the morning breeze. The scattered horizontal clouds were beginning to build into the beautiful cumulus messengers of "angel's tears" as Gus Angermeyer so aptly described rain drops.

We arrived at Post Office Bay, Floreana. The long, white sandy beach offered good swimming. We explored the keg, decorated with the names of vessels that have called at the Bay. Another structure had been built to accomodate the many plaques of ship names which had become too numerous for the keg. Remaining foundations indicated locations of homes of Norwegian colonists who attempted to settle here in 1927. They must have been very disillusioned when they realized the futility of the island paradise to support them. A plaque was made and installed to commemorate the visit of *Pagurus II.* We noticed that the *Super Shrimp* had been here on 1 January. Two days were enjoyed on this island.

About 1545, Brad suggested bending on the Genoa sail so it would be ready for the morning's departure. This activity stimulated our eagerness to begin the 3100 mile voyage to the Marquesas so, by unanimous agreement, we decided all could be readied before dark. At 1700 we were under way and by 1720, we were sailing on a course of 250°, passing Punta Daylight and heading towards a spectacular sunset. Somewhere ahead, the Marquesas, new friends, and new adventures were waiting the arrival of *Pagurus II.*

Irve adjusts to shipboard life on the way to San Diego.

Part of the deck load consists of five-gallon containers of isopropyl alcohol for the Darwin Research Station. Irve takes a break from making baggywrinkle.

Hurricane Fifi brings rain and heavy winds.

The evening sky and waves during Hurricane Fifi are spectacularly ominous.

A porpoise wake seems to direct *Pagurus* toward Boca Grande and Acapulco.

During the intermittent rains, the rainbow indicates the peace we all felt in the port after the storm.

After leaving Acapulco, the aftermath of the hurricane was observed a thousand miles offshore by visits of many birds. Swallows, finches, and hawks are not normally found so far from land.

The Ecuadorian flag was raised when islands of the Galápagos appeared on the horizon.

Volcanic peaks of Isabela Island appear as several islands.

Porpoises escort *Pagurus* into Wreck Bay at San Cristobal Island.

Brad trims Russ's hair as we approach Wreck Bay.

Pinacle Rock as seen from our anchorage at Bartolome Island.

A yellow-crested night heron.

Marine iguanas on Punta Espinosa, Fernandina Island, warm in the sun after feeding on algae beneath the ocean's surface.

A Sally Lightfoot crab displays its brilliant colors.

Marine iguanas swim to and from their feeding areas.

On Tower Island, a male frigate shows its courtship display.

A female frigate shades her young on Tower Island.

A swallow-tail gull with newly hatched chick.

Gus Angermeyer

Karl Angermeyer

Some Residents of Puerto Ayora, Santa Cruz Island

Jackie, Andre, Tui, and Gil De Roy

Waved Albatross, Hood Island.

Russ examining post office barrel at Post Office Bay, Floreana Island.

Brad affixing *Pagurus* plaque.

Pagurus was here!

Tuna provided a welcome change on our passage to the Marquesas.

Cape Martin, Nuka Hiva, announces the end of a beautiful 27-day passage.

Bathing children

Taio Hae, Nuka Hiva, Marquesas

Waiting, watching, and perhaps, dreaming of distant lands.

Returning from the bakery

Atuona, Hiva Oa, Marquesas.

Leaving Traitor's Bay, Hiva Oa.

Hana Vava (Virgin's Bay), Fatu Hiva, Marquesas.

Village children are paid visits by Chief John daily.

Chief John of Hana Vava wears a jacket.

Pamplemousse (grapefruit) were given to us as a farewell gift as we left Hana Vava.

Our anchorage was in 15 fathoms of water in this old crater near Maroe, Huihine.

The road connects Huihine Iti and Huihine Nui as it passes through many small villages.

Nan is ecstatic as we negotiate Passe Teavanui to begin our northward passage to Hawaii.

King Neptune holds court to initiate polywog Nan when we cross the equator.

The equator crossing is celebrated by a cake that Brad baked for the occasion.

A beautiful sunset brought foreboding of a very rough trip from the equator to Hawaii.

After a month in Ala Wai marine, at Honolulu, *Pagurus* was ready to return to the water.

On the way home from Hawaii, a broken forestay was quickly repaired, as though it were a daily occurrence. We had learned to handle emergencies efficiently!

Although our voyage had ended, for many, perhaps such sunsets are yet to be experienced.

CHAPTER VI

TAIO HAE, NUKA HIVA, MARQUESAS

As we passed Punta Daylight, a rain squall bid us a wet farewell to the Galápagos. Then the golden sinking sun lit up the heavens in a splendid array of colors. Its welcome warmth was yet sufficient to dry out the cockpit as we picked up a course of 250°. We made minor adjustments on the large Genoa sail which was billowing out on the port side in the light but steady NNW evening breeze. With the engine off, we were making seven knots in the stillness of a most beautiful and pacific ocean.

It was 17 January, and we were leaving Punta Daylight, Floreana (Charles) Island, Galápagos, Ecuador, at a latitude of 01° 13.5′S, and a longitude of 90° 28.0′W. As we lowered the Ecuadorian flag on the starboard halyard, we thought of our destination, 3100 miles away; Taiao Hae Bay, Nuka Hiva, Marquesas, at a latitude of 08° 57.0′S and a longitude of 140° 05.0′W.

The "Sailing Directions for the Pacific Islands" states:

> The Marquesas Islands. These islands are composed of two fairly distinct groups. They lie in a general northwest and southeast direction between the parallels of 7° 50′S. and 10° 35′S., and the meridians of 138° 25′W. and 140° 50′W. They are all high, of very broken irregular form, and in clear weather they may be seen from a distance of 50 to 60 miles. They are steep-to and only the

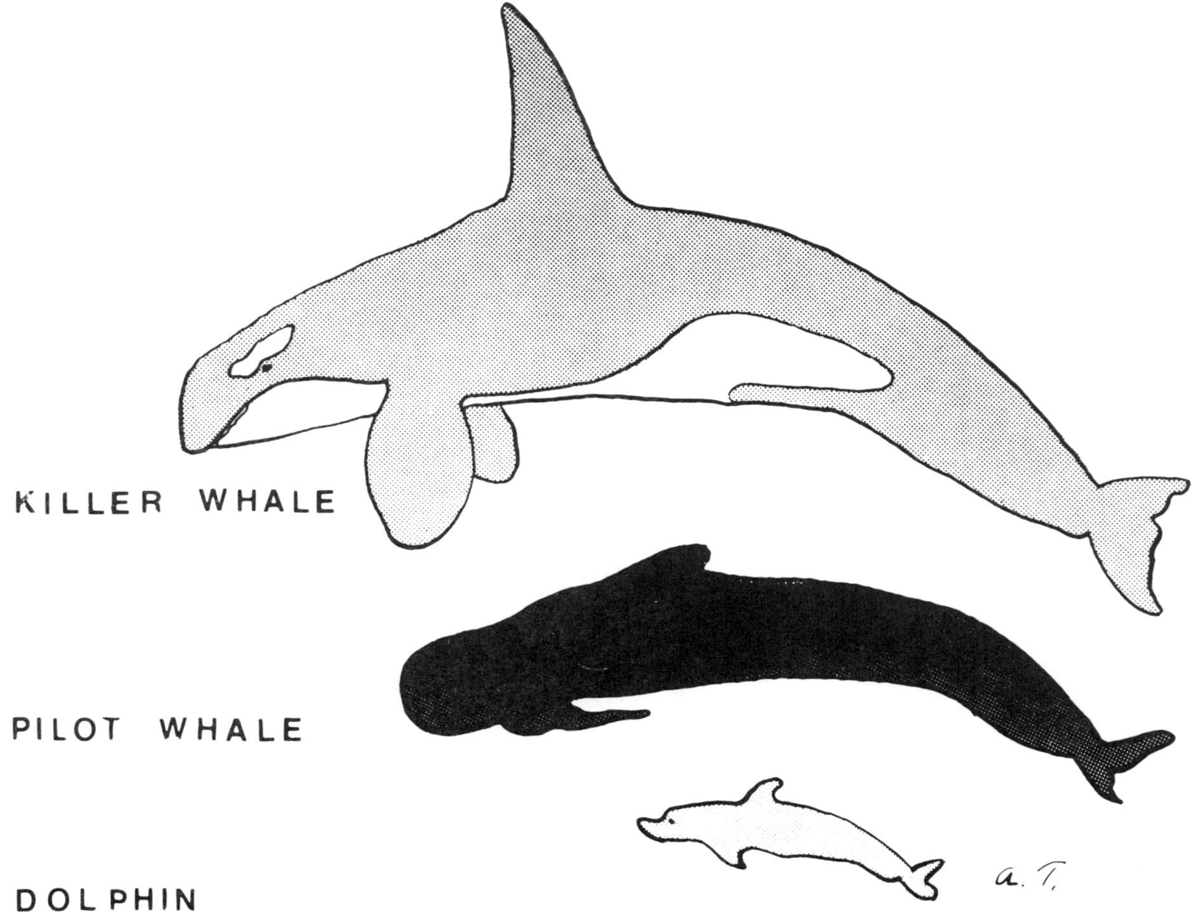
KILLER WHALE
PILOT WHALE
DOLPHIN
a. T.

ordinary precautions are necessary when approaching them.

Landing is not permitted at any of the islands until pratique has been obtained from the Government representative at Taio Hae Bay, Nuka Hiva Island.

Then I was referred to Section 5-41, which described the landmarks, approaches, anchorages, and dangers of the bay. My reading about these exciting, unfamiliar names and places was interrupted when Brad announced that supper was ready. We ate our creamed tuna over mashed potatoes in the cockpit, as the wind vane held *Pagurus II* on course. Brandied coffee served with apricots completed our pleasant meal.

As dusk approached, we talked of the stars that would soon dot the night sky. Since we hadn't thought about the identity of these heavenly bodies for such a long time, we were amazed at the difficulty encountered when retrieving the names from the far recesses of our minds. Our star review continued, and before we realized it, the horizon had blurred and it was too late to obtain a star fix. We really were out of practice, and it was obvious that a sea routine would have to be established again.

We drew straws to determine the watch order; I was first. From 2100 to 2400, I thought the jenny would have to come down because the winds were so light. As Brad and Russ slept soundly, I also thought of the morning of 15 June 1972 when the forty-three-foot schooner *Lucette* was attacked by killer whales and sank in sixty seconds. We were sailing through the same waters. Until reading Dougal Robertson's account of this disaster, I had not thought about such a possibility. Previously, when fishing in a skiff in Monterey Bay, I had been in awe of the size of these animals whenever they approached. I could sense the imminent danger to a small outboard as I started the engine and watched the caudal fins cutting through the water—the tips protruding four to five feet above the surface. But here, in the cockpit of *Pagurus II,* we seemed safe from danger. Many times we have watched them around, but Dougal Robertson certainly must have had the same secure feelings.

Four luminescent streaks heading from the starboard

stern quarter suddenly alerted all my senses, and it felt as though my heart had stopped. The cry to Brad and Russ was stifled in my throat when I heard the squeaks of the dolphins as they approached and dove through the bow waves, only to return and repeat their graceful antics. Although relieved by the dolphin visit, my mind went over the first aid kit, supplies, and lashings of our lifeboat and dinghy. Tomorrow, we would have an emergency drill, and the next day we would have a "man-overboard" drill, I decided.

Our night watches were only three hours, so Brad put on the safety harness and came up at 2400. The wind seemed a little stronger, so the jenny was left flying. Then, after a sound sleep, when I came up at 0600 with coffee, Russ was at the helm, the wind vane holding a good course, and the speed varying from 4.8 to 6.2 knots. After our morning coffee, Russ disappeared below for a short nap, and I was alone in the cockpit. The morning held a promise of a good day as we glided smoothly through the ocean on our way to meet the steady southeast trade winds.

That night our star fix showed that we had only made sixty miles during the last twenty-four hours. We would have to do better than that or we would be lucky to make the Marquesas by Easter! On the night of 19 January the stronger and steady SE winds allowed us to shift the Genoa to the starboard, and we sailed along between nine and ten knots. This was exhilarating! Using Rigel, Achemar, and Diphda, a 1900 star fix placed us at latitude 02° 49.0′S and longitude 96° 31.0′W. During the last twenty-four hours we had covered a distance of 152 miles on a course of 263°. Truly, it had been a good day.

During the next morning, Brad decided to try his luck at fishing. We had a few steel leaders left, so he rigged one up with a spoon and some heavy line. This was deployed off the taffrail and hooked up to a bell. His theory was that if anything were caught, the bell would be pulled over to signal a welcome dinner entrée. When this had been accomplished, our attention was suddenly diverted by a streaking, squeaking, group of dolphins accompanied by several pilot whales. We forgot the fishing as we watched them play at the bow, taking turns to dash past us, dive under, and then sound ahead of us. This was the second time we had observed pilot whales and dolphins performing together. Luckily, they used good judgment, although we once heard a light thump on the boat's bottom, as one errant mammal made contact. Soon more pilot whales joined us until they outnumbered the dolphins. Twenty to thirty of these large animals were all around. I made several 90′ course changes, but they followed suit. They appeared to be about twenty-five feet long, and several times some would jump clear of the water. As long as we traveled at 3.5 knots, they stayed with us, but when we reached 4 knots, they seemed to lag behind. Armed with this observation, I took a course that allowed sufficient speed to leave their area.

At 1230 Brad and Russ were working out the noon sight when the bell suddenly reminded us of the forgotten fishing line. Excitement! We soon had a tuna aboard, filleted, and in the galley waiting for dinner.

Storm petrels darted about the small whitecaps that formed in the afternoon breeze. We saw a pod of about twenty sperm whales going southwest. The closest one passed about 100 feet from us.

We talked of the exciting events of the day during our evening meal of fresh tuna, string beans, coffee, and pudding. Then, the evening stars provided a good position fix before the first night watch was left alone with his thoughts and the safety of two other voyagers below, quietly enfolded in the arms of Morpheus.

At 0400, Brad (who was then on watch) came below and woke me, "Dad, come here to see something interesting." We were surrounded by pilot whales continually surfacing, some as close as three feet away. A glance at the knotmeter showed 2.5 knots. I woke Russ, and we all watched them until they left us at 0530. By this time the wind had stopped altogether, leaving us in a still, mirror sea, as the warming sun rose in the Eastern sky. At noon, I dove off the boat and swam laps around *Pagurus II* through the tranquil ocean surface above the nearest land—miles below. Gentle puffs of wind began to propel *Pagurus II* forward, giving warning that the swimming session was over, so I climbed aboard, feeling refreshed and good from the exercise.

In spite of the seemingly slow progress, the fix on evening stars showed our distance to be ninety-five miles during the last twenty-four hours. We were now more than 600 miles from the Galápagos. To a landlubber, this may be barely more than an hour's jet travel, but to a seaman, this was viewed as one-fifth the distance to Nuka Hiva.

I thought of home and how I wished that Nan could have been with us. The winds were holding steady for longer periods of time, and the wind vane was like having another crewmember, relieving us to read, play backgammon, and take turns planning to outdo the other with gastronomic surprises, created by combining unlikely canned ingredients. I could easily spend an hour's "off-watch" time lying at the bow, watching the parting water as we glided along. I observed thousands of small salps and ctenophores (marine organisms of a jellylike nature) in the water—appearing to separate and form a path for our passage. Even though there were whitecaps, the waters seemed to be flat. The swells were only 2 to 2.5 feet, so we didn't "bury our nose" in them.

I wondered if captains of most small boats were as apprehensive as I. One morning I awoke about 0430 with an "awareness" that something was unusual. We were on a good heel, doing a comfortable nine knots. Somehow, I was aware that Russ had come below to the head, and that Brad had temporarily relieved him. I seldom slept soundly while under way, as I was always checking the duty man to be certain everything was all right. I'm sure Brad and Russ never realized how many times every night I went to the companionway to look up and be comforted by the sight of the dark figure of the helmsman against the star-studded sky. (Often while I was on watch at night, I saw a figure at the hatch—apparently checking me also!!)

Anyway, this time I was a little upset when I found Brad at the helm without his safety harness. Brad said I was being "picky," but I *knew* I was not, and one of my cardinal rules had been broken. When Russ emerged from the head, I informed him that the security of the "ship" was in the hands of the man on watch, and that when I discovered another man on watch without a harness, I didn't approve. I'm sure I was much more abrupt than this account reveals, for Russ didn't answer me, but merely proceeded up the companionway.

When I searched the wet-clothes locker for another harness, I discovered it hadn't been cleaned. I had asked Brad and Russ to clean it two days ago, because something was smelling very foul. To add to my wrath, during the search, something fell out on my foot. I picked it up and found a slimy, rotting, stinking squid! It must have sailed in through the port several days ago.

By this time Brad came below, so I didn't continue my search. I didn't sleep, however, for the rest of Russ's watch, so at 0550 I got up and made coffee. Russ and I silently enjoyed the steaming brew. I decided to talk to both of them later on in the day when I could approach the problem in a calm manner.

A subtle apology and recognition of an awareness of the problem was apparent to me when Russ, later, prepared a breakfast of cornbread, jelly, bacon, and coffee. Sometimes, after tempers flare, actions can speak louder recognitions than words.

The following morning provided a spectacular dawn. The few scattered clouds in the east reflected the sun's light long before it appeared above the horizon. It would be a hot day. We were certainly in the steady SE trades now. The force of the wind varied a little, being lightest at dawn and gradually increasing until about 1100. So far, the passage had been just as people depict sailing; it was ideal. We baked in the oven, had regular meals, and even had snacks with an afternoon cocktail. We read, played backgammon, chess, checkers, and cards. Russ was spending time on working up an outline for writing his thesis, and everyone seemed to be content in the daily routine, now well established.

Sitting at the helm with the wind vane on duty, I enjoyed the feeling of the warm sunshine caressing my skin. No more than a bathing suit was to be required for the remainder of the passage to Nuka Hiva, and even this attire could be abandoned. Brad and Russ were still sleeping, and I let my thoughts wander and reflect.

Yesterday we had opened two jars of jam we had purchased in Puerto Ayora. One was apple, and one was orange.

They were disappointingly alike—both tasted like sugar and no flavor. They were made in South America. Most products come to Puerto Ayora from Guayaquil, and there are very few things made in Ecuador. The olive oil came from Spain. Lucky Strike cigarettes were made in the U.S.A., but stamped "ECUADOR."

I also thought Ecuadorian conservation laws were a bit strange. They really have no way to enforce them. On the Galápagos, it seemed that many scientists have played an important role in developing the conservation laws. Strict attention is given to foreigners, but not to Ecuadorians. The people living on Santa Cruz Island did not seem to be very industrious. They appear to be content with meager "scratchings" for food and the regular flow of tourists to bring in some money. The few stores aren't able to get supplies unless they are on good terms with the owner of the *Pingüino.* Also, many tourists with whom I talked didn't seem delighted about visiting the Galápagos. I was surprised, but many said the only reason they were there was because they wanted a South American tour, and the Galápagos was an included part. Perhaps the tourist trade will have a limited life.

The Ecuadorian government also wants to handle all their own cruise ships, and I believe this would be good for their economy. However, the *Calicuchima, Iguana,* and the *Floreana* are not well maintained. In fact, in 1975, the *Floreana,* which had previously been under Greek registry as the *Lina-A,* sunk midway between Guayaquil and the Galápagos. All hands were saved. The tourists complain about the high price and inferior accommodations aboard these ships.

From discussions with people in Puerto Ayora, it appeared that when a ship like the *Stella Maris* comes in, the Ecuadorians don't want to give permission to land passengers because they believe it will take money away from their own tour ships. They still receive six dollars (U.S.) per person for the Park Service.

I wondered what happened to Señor Ochoa, who was port captain of Puerto Ayora when we had arrived in November. I had heard that he was recalled to the mainland for allowing the passengers from the *Stella Maris* to disembark at Puerto Ayora. I also heard that after the *Stella Maris* had discharged her passengers at Guayaquil, another group of passengers boarded. Then, she came back to Tagus Cove, Isabela Island, where she was met by an Ecuadorian navy ship and asked to leave the waters of the Galápagos. Certainly, there must be much more to the political situation than meets the eye. Perhaps some day the Ecuadorians will have a beautiful passenger ship like the *Stella Maris* that can bring passengers to the islands first class. For conservation, however, the numbers of people visiting the islands should somehow be controlled. I believe that the designation of specific sites for visitation will restrict the areas for tourists, allow people from all over the world to see these fascinating islands, their fauna and flora, and minimize the disruption of breeding grounds.

My musings were disrupted when the lower bracket of the steering vane suddenly broke. Since this was a three-quarter-inch aluminum tubing holding the canvas vane itself, the problem wasn't too serious. I maintained the helm while Brad and Russ cut off an eight-inch piece of plastic hose. Then they inserted the tubing and taped it as tightly as possible. This "splint" arrangement allowed a little "play," but it seemed to hold. Well, I thought, when we get to Papeete, I'll have to see if I can get the bracket repaired more firmly, perhaps, I can get a smaller piece of aluminum tubing, insert it *inside* the existing one, and then drill holes to bolt them together. For now, the course was steered, and the temporarily repaired wind vane reset. The course held.

Today was also a "repair day" for several sail slides that had come unsewn. Next, we had to lower the Genoa to resew two jib snaps that were loosening. If these weren't strengthened, a heavy wind might cause several to give way and serious problems could follow.

The fourth item requiring attention was the automatic bilge pump. It seemed to be going on frequently. We removed the companionway ladder and cleaned behind it, finding a piece of cornbread, a banana, and three spoons! Then, I crawled into the engine room to tighten the stuffing box. Realizing that this had not been done since Acapulco, I decided to check it more often. At any rate, the slow influx of water was now checked. Next, the battery water level was inspected and filled where necessary. Since we hadn't used the engine for a week, the stillness was shattered by the diesel engine for an hour. A warm feeling of accomplishment accompanied dinner that evening.

The celestial fix using Aldebaran, Achenar, and Diphda showed that we had made a headway of 152 miles during the last twenty-four hours. As night approached, the nearly full moon lit up the cloudless sky. We were more than 1000 miles from the Galápagos and still east of the longitude of San Francisco. In a few days we would pass the halfway mark. Squid and flying fish occasionally flopped on the deck as we enjoyed the comfortable warm, clean air and munched popcorn which Brad had made for this relaxing time of the day.

27 January. I thought it was 0600 when I arose—it was 0700. Russ had taken an extra hour's watch so we could set our watches back for a time-zone change. Feeling guilty but refreshed, I received information about the preceding four hours and took the helm while Russ went below to chart the night's run. Our steering vane hadn't been holding a steady compass course for the last two days because the winds had been changing from the S to SSE. During the various watches, the compass course varied from as low as 230° to as high as 245°, but the average was 240°, so we were making our desired track with an average speed of 6.8 knots. The ocean swells were whitecapped and were also changing from the NW to W, giving us choppy seas, but following swells, thus requiring more attention to the helm.

The blue skies with scattered cumulus clouds added to the changing seascape, and the days were warm. Flying fish and squid were always on the decks in the morning. These provided good bait for fishing. Alone with my thoughts, it was interesting to realize that we had progressed more than 1,000 miles west of the Galápagos. We were at 108° W, and another 1,000 miles would be traversed before we would reach the longitude of San Francisco.

The cumulus clouds appeared to coalesce, and as the day progressed, the sky became overcast and dreary, providing us with a wetting drizzle. We were able to soap down and rinse off in the light rain, but it took a little longer than the sudden envigorating cloudbursts to which we had become accustomed. I think Russ got carried away with the clean feeling of the fresh water, because when he finally came up for his watch at 1400, he had shaved off his beard. I hardly recognized him!

Our evening star sights gave us a good position fix at latitude 06° 54.0′S and longitude 109° 13.0′W. Our set was heading us in the right direction for the Marquesas.

The next morning our star position fix was latitude 07° 41.0′S and longitude 111° 27.0′W. Over the last eighteen hours we made 151 miles of a true course of 251°. Our average speed was 8.39 miles an hour. Our speed was really greater than it seemed. But on the morning of 30 January, the wind became light, and we slowed to four knots. Two tropic birds that had been with us for several days still circled occasionally, and at times would alight and sit on the water. Several storm petrels could always be identified.

Today we were to pass the halfway mark. I thought about the sharp contrast of this voyage to that we experienced from San Diego to the Galápagos. I recalled the apprehension felt when leaving Floreana, but so far this voyage had been idealistic. We had been cooking full meals, observing the birds, flying fish, squid, reading, and playing backgammon. The boat had really required little attention. Other than sewing an occasional sail slide or jib snap on the Genoa, our "shipboard life" was truly a lazy one.

Brad made a cake for our "halfway" celebration, and I fixed Polynesian curried chicken with the last of our rice. Janet McNabb had given me a can of Mandarin oranges, so I added this, as well as pineapple chunks, to the curried chicken. We were out of coconuts and coconut milk, however, and I think this dish has a better flavor with it. Oh well, *c'est la vie*. The table was set below with cloth, candles, and artificial flowers. We had a pleasant celebration dinner before sighting the evening stars, and the wind vane kept us on a smooth course.

Canopus, Aldebaran, and Betelgeuse gave us a perfect position fix at 1900. We were at latitude 08° 46.0′S and longitude 114° 24.0′W. Since Taio Hae is at a latitude of 08° 57.0′S, we were really at a proper latitude for the landfall. The steering vane wouldn't hold much more than 260°, so I decided to steer the helm for a while. Although we were on a broad reach with following seas, our speed was a good 6.5 knots. Considering the fact that we were taking a NW set, we could steer a desired course. If we could have steered 210°, our speed would have been eight knots, but I decided to sacrifice the speed for the course. For those below, the course and speed were very conducive to restful sleep.

Alone at the helm, with Brad and Russ sleeping peacefully below, I thought about last evening's sunset. It was clearly the most beautiful of the entire voyage. The reds and oranges of scattered clouds against the dark blue sky provided us with a spectacle of nature which no camera could truthfully record.

We made 131 miles that day. The following seas created steering problems, and the wind vane didn't hold as well with the light winds. It was discovered that the wind vane could be set on a course and the wheel *not* secured. Then, by just moving the wheel no more than one spoke in either direction, an average course of 255° could be maintained. As I watched a spectacular dawn on the first of February, I thought that we should sight the Marquesas in nine days. Then, I decided to make a calendar so we could anticipate the landfall. Truly we were on the "short" end of the trip now. The anticipation of a landfall is really a morale booster to the sailor at sea. I could only imagine what Tao Hae would be like. The "Pilot" says that fresh fruit, vegetables, meat, and good water are available. I could only guess as to the "friendliness" of the 150 people that live there. Time would tell. Now, the sun was just appearing above the eastern horizon and a brilliant golden color appeared in the scattered horizontal clouds. Dawn was here, and so was Brad's watch.

Brad came up with coffee just as the fading gold in the sky was replaced by the beautiful sun and the blue sky of day. I wrote in my journal: "I believe Russ is keeping a journal and making daily entries, as I am. Brad isn't, but he tries to write Lynn every day, so I guess it is the same thing."

In analyzing the creature comforts of transoceanic passages, I now realize that boom gallows for the main and mizzen would have been most helpful. Then, another dodger could have been built from the main boom gallows to protect the cockpit area. The helmsman could then be on duty and protected from unwanted downpours. Even though the temperature has been most agreeable, the rains are chilly and a wet body is not always comfortable. Also, on the design of *Pagurus II,* blocks should be placed on the deck aft of the cockpit so the water coming aboard at the bow and running aft doesn't meet on the poop deck and come into the cockpit. Perhaps the wheel house could be redesigned to curve around the lazarette to form a more comfortable place to sit. Wider seats in the cockpit would also aid. And, the cushions for use under way should not be canvas, because they seldom dry when they become saturated with salt water. Then, of course, one is always sitting in a damp (or wet) spot!

My electrical switchboard panel is located beneath the port lazarette. I believe this would be better if it had been located below. We hadn't had any trouble on this voyage, but in the past, a wave had wetted the various terminals, causing difficulties.

As for cruising, a saltwater tap in the galley sink would have been very valuable. Although we usually washed our dishes on the leeward deck, this is not the safest procedure. During inclement weather, a bucketful of sea water carried below would often spill, and the wet carpets would require several days to dry. A saltwater tap in the head sink would, likewise, be a useful convenience and save our fresh water. In the South Pacific, however, we didn't find fresh water to be a problem for washing, because there was always an abundance of rain squalls. Our water tanks held 100 gallons, and we carried 20 gallons on deck. This proved to be more than adequate.

Larger fuel tanks would have been an asset. There were times (through the doldrums) when I might have powered, had I not been so conservation-minded. Although a deck-load of fifty gallons of diesel was carried, our two tanks held only a total of fifty-eight gallons. Not knowing the availability of diesel in various ports-of-call, made me hesitate to use our fuel aboard.

We had been making several sail changes during the past few days, and on 4 February the wind seemed to be steady again from the ESE. We were now 975 miles from our landfall. It seemed to be good to realize that two-thirds the distance had been traversed. Our calendar now indicated "six days" but it might be more like eight, I thought at the time. The weather had been good, and *Pagurus II* had behaved well. We all adjusted well to the voyage, and had an inner feeling of accomplishment to realize that our location could be pinpointed in this vast Pacific Ocean. Proof of this confidence would be withheld until we reached our landfall!

On 12 February our position fix placed us 240 miles from the Marquesas. We were already two days beyond my projected arrival, which was made more than a week before, but the proximity to land lifted our spirits. The night before we covered sixty-two miles with good strong winds, so now we had 180 miles to go. I hoped the wind would hold, so we could spot Ua Huka before nightfall. This island is thirty miles east of Nuka Hiva. Then, we could proceed slowly during the night and go into Taio Hae, Nuka Hiva, in the morning.

Last night we had had very heavy seas and strong winds. We were running with the main and mizzen, wing and wing during the day, so we could head west. But with nightfall approaching, I put both sails on the starboard to stabilize the boat for comfort of the crew below and for ease of the watch on deck, as we headed 245° Each of us, during our watches, caught breaking seas flying into the cockpit. It was as if the South Pacific didn't want us to become too complacent. Without this, our 3100 miles from the Galápagos would have been almost "uneventful."

The first pooping wave hit as Brad was on watch and Russ and I were working out the evening's celestial fix. About 1930 there was a thud, shudder, and a splash, as torrents of water came below through the companionway. We heard Brad glugging a few choice words, as gallons of water receded in the self-draining cockpit. After checking to see that all was well, I tossed him a foul-weather jacket, closed the hatch, and Russ and I continued on the solution of our navigational problem. A little consolation occurred when we completed our calculations. The position fix was a classic textbook example of three lines intersecting at exactly the same point. At 1900 we were at latitude 08° 22.0'S and longitude 138° 57.0'W—seventy-two miles to Nuka Hiva plus about seven miles to the bay from Cape Martin. According to the "Sailing Directions," it was necessary that we clear pratique at Nuka Hiva.

As fate would have it, the wind dropped, and our night's run was only 11.5 miles. At 0610, Russ woke us with "Land Ho." What a reassuring and exciting message. There was Oa Huka, about twenty or thirty miles away. How impressive it was to actually see the small 7.5-mile-long island come into view after 3,100 miles of ocean.

Realizing that our goal was only seventy miles away, I started the engine to replace the power from the insufficient breeze. By 0800 we could see the tops of Nuka Hiva's mountainous peaks, as we continued under a leaden sky.

At 1700 on 14 February we dropped anchor on a sandy bottom at Baie Taio Hae, Nuka Hiva and cut the engine. Also anchored here were the *Catapha* from San Francisco and the *Maris* from Australia.

The stillness of the anchorage made us dizzy after the twenty-seven-day passage. By invitation, we went aboard the *Catapha* and met David White, captain, as well as Ian Kieman, captain of the *Maris*. Strangely enough, Ian's crewman was Mark Fraser from Half-Moon Bay, California. After much wine and talking, we returned to *Pagurus II* for a Spam sandwich, as we hadn't eaten all day. Then, we fell into sound, dreamless sleeps until morning.

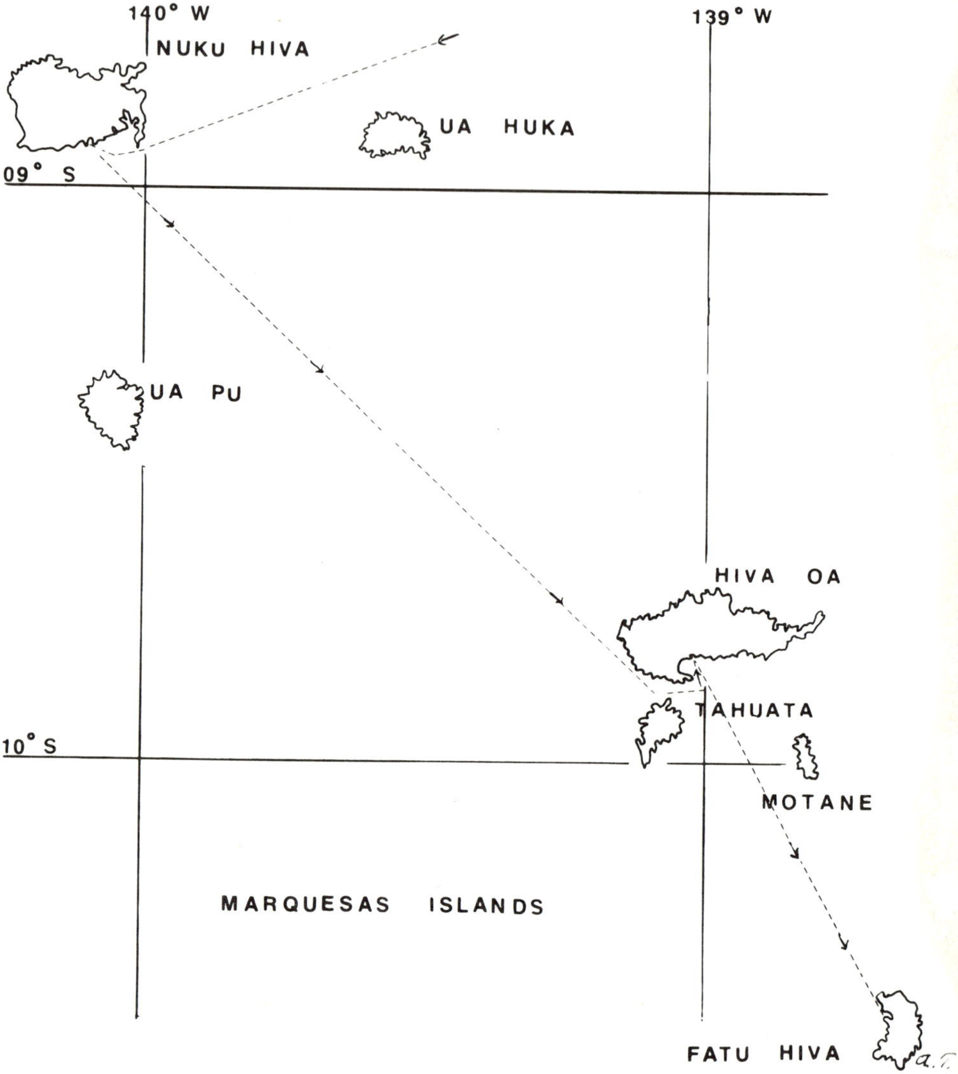
140° W
139° W
NUKU HIVA
UA HUKA
09° S
UA PU
HIVA OA
TAHUATA
10° S
MOTANE
MARQUESAS ISLANDS
FATU HIVA

CHAPTER VII

A MARQUESAN ADVENTURE

Awaking at 0630, we dove off the boat and were refreshed by swim in the clear, cool water. Then, letters were gathered for mailing and the "ship's papers" were placed in a packet for presentation to the officials. The gendarmes were very friendly, and the *douanier* was most helpful and patient with my terrible French. All papers and visas were in order, and we were welcomed to the Marquesas.

What a contrast were these lush green mountains when compared to the cactus-strewn bare lava landscape of the Galápagos. There were three stores in town: Maurice's Trading Post, a rather modern looking store, and another smaller one. Behind one of the stores was the bakery, where we purchased two loaves of French bread for two francs each. Then we purchased two chickens, butter, rum, Scotch, oil, soap powder, rice, and two ice cream cones each. Maurice took us up the mountain where we picked ten grapefruit (pamplemousse). They must have had a combined weight of forty pounds!

By the time we rowed back to the boat we were ready for another refreshing swim. Then Brad was making mai-tais for beginning our celebration and I was in the head when I heard a commotion. Russ excitedly called, "Al, there is a savage here." As soon as possible I emerged from the head and saw a wooly head peering through the galley port. Brad, Russ, and the stranger appeared to be ill at ease, and I wasn't certain what problems were facing us.

"Bonjour, mon ami; je suis capitaine de bateau. Je m'appele Al." The apparition answered: *"Je m'appele Max—Maxime Tehuitua. Maurice dit moi que vous voulez des bananes. J'ai des bananes."*

Max was standing in his outrigger still peering through the port with a perplexed expression because Brad and Russ had not understood that his intentions were friendly, and he was merely offering us bananas, mangoes, pamplemousse, and papaya. I quickly invited him aboard to make amends, lest he think we were the unfriendly ones. He didn't drink wine or liquor, so he had a hastily prepared citronade, some bread, and cheese. I showed him through the boat, and he was most interested in the way the food was stowed and the amount of supplies we had aboard. I gave him several cans of corned beef as he left, saying that he expected us at his house at 1500. He would have plenty of fresh fruit for us.

A hot two-mile hike up the mountain brought us to Max's house. We met Junie, his wife, and three of their six children. Their eldest boy was eighteen and a radio technician in Papeete. Junie explained that six was enough, and then both she and Max laughed. We were seated at a table in their "kitchen house" (which was separate from the sleeping quarters). Max showed us pictures of all his American friends and their yachts. Then, we were given mangoes, bread, breadfruit, and fish to eat.

We learned that Max was from Huahine, a Tahitian island near Papeete. His wife was a Marquesan, and a descendent from a once-powerful chief. She explained that she was a princess. Her nearly six-foot height contrasted to Max's five foot, seven inches. She giggled all the time, showing her best grin, and not at all conscious of every other missing tooth.

Junie had been ironing on the floor of the cookhouse–dining area when we arrived. The iron had a little tank on back and was heated with petrol. Bowls for their meals were washed and put upside down on a shelf at the window to dry. There were no doors or closing windows.

They took us next door to see the living room and three bedrooms, which were in a separate building. Max showed us the completed foundations for the "big house" he was building. He said that it would be completed in one year. Nearby,

Junie's parents lived in another small house.

Time passed quickly, and as it was getting late, I explained that we should return to the boat while it was still light enough to see the trail and then the road back to the beach. Max stood up, took his machete, and we followed him down a trail where he climbed a tall coconut tree, and cut about ten coconuts, letting them fall to the ground. Next, he made a spike and husked them for us. By the time we began climbing back down the mountain, we were carrying a large sack of coconuts, a big plastic pan with mangoes, papayas, and breadfruit, and two stalks of bananas. He wouldn't take a single franc for all this wonderful fruit. He explained that he had fruit, and we had none; now we have fruit, and he still has more.

Max's generosity of thought and time was further shown by his desire to take us spear fishing in the morning. He carefully explained that specific fish were good to eat when caught in one cove, but in another cove, the identical fish was taboo.

It was quite late by the time we returned to *Pagurus II* that evening, and I was frying chicken for dinner when Dave White rowed over to ask us to join them for a "bugaloo" ashore. "What is that?" I asked. "Everyone brings a fish and some wine; we build a fire and have a party," he explained.

"I have dinner almost ready, so we had better eat aboard. Then, I'll make a rum punch, some snacks, and we'll join you ashore later."

"Good," he said, as he shoved off for *Catapha* to pick up his crewmen.

One meets many interesting people of varied backgrounds, interests, and abilities when cruising. The crews from *Maris* and *Catapha* we had already met. We were then introduced to Rio and Charles. Rio was a girl from Redondo Beach, California. She had been traveling aboard a twenty-seven-foot boat that arrived at Taio Hae three weeks previously. For some reason she left (or was left) here when the boat sailed away and had been staying in a room of a little cottage ashore. Charles explained that he left his home in Argentina two years ago because of the "unfavorable" political situation there. He said that he had just been traveling throughout the

South Pacific during this time. Charles appeared to be well educated and have some means of support. He spoke English and French, as well as his native Spanish.

Then, we were introduced to two Marquesans, whose names were uncertain! Apparently, they invited themselves to the "bugaloo" and had been drinking for some time. Their contribution to the fare was several large breadfruit, which they were roasting in the fire. Beer and Algerian wine purchased from Maurice's Trading Post were consumed with gusto.

One of the Marquesans seemed to single me out for his conversation partner for the evening. He was friendly, but yet there was something about him that made me unsure of his character. Not wanting to offend him, I listened. He was very definite in his likes and dislikes. He kept telling me that the Marquesans were good and "Tahitians no good." Obvious hatred was venomous when he said, "No good." He said that Tahitians were mean. "One drank beer and then cut my arm with his machete." There was a large, but healed scar on his arm. He didn't like America, Italy, or France because they were "too big." However, he had never been to these countries. He didn't like the *"bleau, blanche, et rouge"* and thought Marquesas should have its own flag. I began to sense his feeling for his land and his people, but conversation via a combination of French, English, and Marquees became more negative with every beer. He often repeated, "papaloo . . . papaloo," then he would say, "mon ami" and give me a hug.

He wanted me to go up to Maurice's with him to get some more beer, but I told him, "No, I prefer to remain here in this warm circle of firelight." Finally, he left to walk up to the store, so I moved to another spot to converse with the other people.

When my Marquesan "friend" returned, he sat on the fringe of the group and kept yelling for me to come over and talk to him. It perturbed me to be rude, but I tried to appear not to hear him, as our conversation had become so negative and repetitious. About 2330 when we were leaving, he took one of our windbreakers that Russ had brought ashore, and he put it on. He said that he would meet us on the beach in the morning. I explained that we needed the *chemise,* but he was quite resistant and refused to remove it. Finally, I told Russ to forget it. Two days later, a gendarme brought our jacket and apologized. I wonder how the gendarme learned of the incident, for I certainly did not report it.

The next morning (after the "bugaloo") the other boats left for Baie Tae Pae and took several Marquesans with them to join in a local celebration there. We would have liked to go, but Max was coming to go diving for fish and lobster with us. He arrived about 0800. At 0810, Max's outrigger shoved off *Pagurus II.* Max and Russ each took a rowing oar and Brad steered from the stern with a short paddle. I sat at the bow, and Max laughed as he pointed at me saying, *"Capitaine."*

While diving among various corals, I noticed two quite different sea urchins. One was cream colored with tan longitudinal bands and one equatorial band. Another was all tan. When I brought one to the surface for closer examination, I was surprised to find that the bands weren't tan at all, but the typical purple of many sea urchins. Some tube feet were purple while others were cream colored.

The variety of colors of the many kinds of fishes never ceases to be amazing. Tiny vivid blue ones continued to dart about in schools, all moving in unison. Many medium-sized bright yellow or shining gold fish darted about independently. Another, about twenty centimeters long was green with yellow and blue markings. Its pectoral fins were also green and yellow with bright blue margins. Typical Sheepsheads and Moorish Idols, were everywhere. Parrot fish were nibbling at the coral heads. It was as if we were swimming in a tropical fish aquarium.

At the second diving spot, between Sentinel East and Nuka Hiva, Max gave Brad his long speargun and said, "Go fish." Brad later came up with (what I thought was) a beautiful red rock fish exactly like the one Max had speared in the first cove. "Uh—no good—poison," exclaimed Max, as he removed the fish and threw it back. "I show," he said as he took the spear and dove into the crystal clear water. Soon he was back with a red-colored fish about thirty-one centimeters long. "Theese goooood," he told us. To me, it looked like the one Brad had caught. Then, as Brad and Russ were off looking for more of these edible fish, Max dove and brought up a large conch which he said was also very good to eat.

After every dive, we had a mango from a basket provided by our new friend. Max didn't eat any, saying, "No, you enjoy; my house much fruit, you no have fruit, you eat." We found this unselfish philosophy and joy of sharing throughout the Marquesas. Life here is simple and the greatest pleasure was just in living every moment of every day. Life is precious and beautiful when one's attitude is in tune with nature.

It was surprising how easily the outrigger glided through the choppy afternoon waters as we opposed the current's direction when returning to *Pagurus II.* Although the waters were warm, after diving all afternoon the breeze was chilling. Back aboard, Max used my binoculars to scan the shoreline. When I asked him what he saw, he replied, *"Mon petit enfant sur le sable."* He said that his baby goes down to the beach all the time. His youngest was two years old. When asked if he was worried, he said, "Oh no, everyone watch, and baby goes all time."

Max told us of Tae Oa where there was a big waterfall we should see. He said we would enjoy a visit: "One day, gooood—two days gooood—more days, no goood. In Tae Oa," he explained, were "seese homme—no vahine. All homme friends—all homme good—if one vahine come, no gooood—homme no friend no more. Ua Po—many vahini." We can see the tiny island of Ua Po about twenty miles away from our anchorage by looking directly through the entrance to Baie Taio Hae.

After our visit, I returned with Max to help him take the outrigger ashore. Then, I carried the oars to his house as he carried his diving gear and several fish. On the way back, I picked a breadfruit for supper and met Brad and Russ at Maurice's Trading Post. It was closed, and had not been open for a day. Although the hours of business are not always consistent, this seemed unusual, so I walked around to the back. Through the window, I could see Maurice. His usual boxer trunks were the only clothing he wore, to partially cover his big, rather rotund body. He was lying on the counter on his back, with his head propped on a stack of magazines, reading. After gaining his attention by rapping on the window, I

Maxime shows Al his outrigger at Taio Hae, Nuka Hiva, Marquesas.

asked him what time he would open. Directing us to a back door, he admitted us, and we purchased our needs, placing them in our "shopping bucket."

The next day was a work day. The steering vane rudder was removed, scraped, cleaned, and painted. The bottom was cleaned of its lush barnacle growth, and the doghouse sides were sanded and varnished. We even found time to row ashore and do our laundry, using buckets and fresh water.

The *Maris* and *Catapha* had gone out sailing for the afternoon. Ian and Dave wanted to take pictures of their boats under full canvas. They had been invited for dinner aboard *Pagurus II* that evening, and we hoped that we would be ready. As I glanced out to sea to watch the two boats leaving the bay, I noticed a small boat coming in. My excitement proved to be justified when looking through the binoculars. It was the *Super Shrimp.* This was the eighteen-footer from England that we had seen in the Galápagos. She had a red patch on the mainsail. It took her four hours of tacking to arrive at an anchorage not far from *Pagurus II.* We excitedly met Shane and Iris with an arrival gift of mangoes, bananas, pamplemousse, a coconut, a loaf of bread, and some much appreciated cold beer. Naturally, they were invited to dinner so we could all hear about their forty-five-day passage from Floreana, Galápagos.

I fixed curried chicken, using pineapple and fresh coconut milk. We cooked four cups of rice and had a couple gallons of wine. Nine of us spent an enjoyable evening swapping tales. In the early morning, three dinghies shoved off to their respective boats—but Mark and Dave chose to dive in the warm water and swim.

Mail came in the next day, so we walked to the post office to exchange the letters we had written for the news from home. We then walked to the west side of the village to inspect the two churches. We met the bishop of the Marquesas who told us that the present population of the island of Nuka Hiva was 1,400 and the small island of Ua Po had 1,700 people. The old Catholic church had been built in 1857, and a new one was being constructed. When I asked if any of the beautiful stained glass windows in the old church were going to be used in the new one, he told me, "No. The new

Iris Derungs and Shane Acton enjoy their first taste of breadfruit after a 45-day passage from the Galápagos aboard the 18-foot *Super Shrimp* (in the background).

church," he explained, "will be built in the Marquesan style with no closed windows." All the teachers in the Marquesas are from France, and no local people were hired as teachers. The bishop showed us the missals which were written in French and in Marquees. He had been here for four years and thought the language was not difficult to learn.

Walking back to the village, we passed the only available restaurant, where Russ had ordered lunch for us earlier in the morning. Cut up cold roast beef, onion, and tomato in an oil and vinegar dressing was served as a salad. Next, we had Hinano beer, noodles, and thick, rare, steak that had been cooked in butter. There was enough for six people! After this, a sliced tomato and onion salad was brought. We thought this was sufficient, but a desert of bananas fried in butter and sprinkled with sugar was brought on a large platter. Three satiated individuals left the little restaurant!

On the road we met Kirk, Mark, Martin, and Mike. Karl built his ferrocement boat, *Dulcinea,* and was from Vancouver, Canada. They arrived in Taio Hae the day before. Now, they were interested in obtaining a woodcarving, and since we were on our way there, they joined us. We introduced them to Manu. Then, I selected a large chunk of rosewood and asked Manu if he could carve a Tiki for me from this wood. He said that he would do it for 1,800 francs, and it could be done in three days. This was agreeable, because we were leaving in the morning to visit two other anchorages, but would return to pick up our passports.

Tai Pae Vaii is a small settlement to the east of Taio Hae. It has been described by Herman Melville. Its small sandy beach hasn't changed, but the people have. It is difficult to believe that little more than forty years ago, the tribe living here was cannibalistic. Melville has also described the eating habits of this tribe.

We found Baie Tao Oa to be much more interesting. It was tricky to find a suitable anchorage here, as it is hidden from view when at sea. The strong winds and rough seas made me very apprehensive as we approached the rocky shore. With sails furled and engine on, I first mistook Ua Uka for Hakatea, but obviously, there was no anchorage here, so *Pagurus II* headed out to sea again and entered a narrow passage between rocky crags. Then, the Bai L'Anse Hakatea opened up, and the water was calm. Anchor was dropped on a good mud and volcanic sand bottom in eighteen feet of crystal clear water.

The Sailing Directions for the Pacific Island, Volume III uses different spellings than various charts of the area, but section 5-46 states:

> TAI OA BAY, located close west of Ua Uka Bay, contains coves; Haka Tea in its northeastern part and Haka Ui in its northwestern part. At the entrance the western side is formed by lofty and perpendicular rocks, and the eastern side is a long rocky point with slight indentation and the sea breaks heavily along it.
>
> The depth in the entrance is about 36.6 m (20 fm) over a bottom of fine sand and clay.
>
> The bay is only suitable for small vessels, as the swinging room is very limited, and violent squalls come down the valley. Haka Tea extends about 800 yards in a northeasterly direction and is about 400 to 500 yards wide. At the head there is a sandy beach behind which there is a green flat. The depths in the entrance to the cove are 12.8 m (7 fm) to 14.6 m (8 fm) but decrease gradually toward the head. Vessels anchor with the eastern point of the cove bearing 300°.

We lunched on French bread, cheese, tomatoes, avocados, and mangoes before rowing ashore to explore the ravine and jungle of the tropical forest that lay before us. Wild chickens flew among the trees. The narrow trail was obliterated when we came to the large boulder-strewn stream bed. It was late in the afternoon when we returned to the beach to refresh ourselves with fresh coconut milk and a swim.

We were gathering shells when a small canoe with two men landed on the beach. Benwar and Gabrielle were from the village of L'Anse, Hakaui. I recalled Max telling us that six people lived in this village, which was about a half-hour's walk from where we were anchored. The men were friendly but reserved as they continually asked if we had any *"fusée."* Then they would converse in Marquees, and ask the question again in French. When all else failed, communication was effected with sign language. They wanted to know if we had any guns and shells aboard. Disappointment was obvious when they learned that we did not. Many feral goats lived in the hills, and they wanted to shoot some.

Benwar and Gabrielle gladly accepted my invitation aboard where we served tea, crackers, and jam. They expressed amazement at the various furnishings of the boat and wanted to see every nook and cranny. Perhaps they were still looking for "fusée"! It was at this time the *Dulcinea* entered the pass to anchor near *Pagurus II.* I introduced the crew to Benwar and Gabrielle, who immediately began asking about "fusée." When I explained the request to Karl, two guns were presented for Benwar's inspection. Apparently the nine millimeter guns were too large to suit his needs, because Benwar merely shook his head and said, "No good."

Benwar asked if we would like to go for a freshwater swim and promised that we would have time to return before dark. Other than showering in rainstorms, we hadn't had a freshwater shower since Acapulco, and the thought of swimming in a fresh water lagoon was overpowering. Martin Turner, Mark Lemon, Mike O'Conner, Berry —, and Karl Begoich from the *Dulcinea* joined Brad, Russ, and me as we followed our two guides. After rowing to the rocks at the west side of Baie Hakatea, we landed the dinghies and began hiking on a trail across the volcanic, rocky area at the shore and then up a steep path through thorn bush. We were hiking over a peninsula separating the two coves. The trail suddenly ended at a seventy-foot cliff overlooking L'Anse Hakaui, the village of which we had heard contained six men—all good friends!

I recalled the words of the *Sailing Directions* and my few hurried glances of the area as we had arrived earlier, but the spectacular beauty of L'Anse Hakaui hadn't registered. Standing here at the cliffs forming the eastern side of a long beach, I could see the almost vertical cliffs rising 2000 feet, forming a backdrop of a lush green valley to the north and extending to the south and into the clear, blue Pacific. Coconut trees lined the base of this volcanic wall. A freshwater stream flowed down the valley, but it could only be im-

L'Anse Hakaui, in Baie Tai Oa, Nuka Hiva, Marquesas.

agined, at this time, because it was hidden by the tropical growth, adding dark green color to the picture. A long sandbar separated a large freshwater lagoon from the salt water of the bay, and this lagoon was also ringed with coconut trees. On the eastern side of the beach, a wide, shallow stream overflowed from the lagoon. Closer inspection revealed about five structures situated around the lagoon, and we could see several horses grazing on a flat grassy area. An old but much used canoe painted turquoise blue rested on the sandbar. In the warm evening air, our little group of barefooted visitors, outnumbering the population of the village, carefully followed the guides down to the sandy beach and enjoyed the gift of a delightful, cool, freshwater swim before bidding *bonne nuit.*

From our anchorage, the mountains to the east hid the rising sun, and it was nearly 0800 when I woke. Pamplemousse, coffee, and bananas fried in butter gave us a start for what was to be a memorable day. We were to meet the other villagers, and Benwar was going to guide us to a spectacular waterfall about four kilometers up the valley.

Joined by *Dulcinea's* crew, we rowed around to Hakaui and beached the dinghies. Barry didn't join us, as he was going goat hunting. Benwar met us and pointed to his two houses. Proudly, he explained that he also owned some land, two pigs, and a horse. "Other horses roam the hillsides and are occasionally used to haul copra, but I own one." His father, Daniel, the chief of the tribe, came out to meet us. When I first saw him I was startled. He looked exactly like Manu, the woodcarver in Taio Hae, except that he wasn't deformed and crippled by leprosy. A turban around his head, a yellow shirt, and shorts covered his stocky frame, and he greeted us with, "Hello, I'm Daniel," as he knew a little English. When I explained about his likeness to the woodcarver, he asked me if I would please take greetings to his twin brother, Manu. I learned, also, that Junie (Max's wife) was his sister, and there were fourteen children in his family. Except for the immigrants, then, most inhabitants of Nuka Hiva appear to be related in some way.

As we waited for the "right time" we ate some fruit off a tree Daniel said was *"buisson."* The fruit was in clusters, purplish, and looked like elongated cherries. They were a bit bitter, but good. We examined his banana trees and tapioca plants. Then we had a good game of volleyball, and Daniel demonstrated his physical condition by playing a powerful game.

Benwar announced departure time by arriving astride a saddled horse, leading a barebacked mount with a rope halter. Pointing at me, he commanded, "You ride." Wearing only tennis shoes and a bathing suit, I was soon sitting on a most uncomfortable vertebral ridge. When I was atop the skinny beast, Benwar shouted to the others, *"Suivrez moi,"* as everyone laughed, fell in behind, and began walking on the trail beside the stream.

At first, the trail resembled a road built of stones and about ten feet wide. We passed stone walls and remnants of stone foundations, which served as mute evidence of a once-large village. Benwar told me that his grandfather owned all this land of the entire valley. Several stone tikis lined some of the stone walls. As we were now well ahead of the hikers, we dismounted at an old deteriorating church. It was more than 130 years since this church had been built. There were holes in the rotting floor, but the altar still contained candelabra, and the cabinets held vestiments and hymnals which could never be used again. Pictures, some of which had curled out of their frames, still adorned the walls of this worship house that must have, at one time, served as a transition between the thoughts of the "savage" and the "civilized." Benwar explained that the large village began to decline soon after the French assumed control of the Marquesas. They took the school away and put it in Taio Hae, thus consolidating all the children from various tribes. Many children didn't return to their native villages. Also, his father, Daniel, had told him that slave traders used to come in and take people away.

By this time, the hikers had caught up with us and wanted to see the church, but Benwar said, "No time," and off we started. Continuing along the narrowing trail beneath thornbush, mango, banana, and coconut trees, we noticed two more stone tikis on the stone wall. Soon we reached a clearing and Benwar suddenly kicked his horse and took off down the grassy slope. Naturally, my beast had to follow, not

This trail provided easy hiking on Hiva Oa.

wanting to be left behind. I not only held onto the rope bridle, but my stabilization was assisted by a firm grip on a fistful of mane! The next thing I knew, there was a third horse; Benwar's lasso had firmly encircled its neck. Not understanding his reasons, but following directions, I mounted the new animal, controlled it with the lasso, and held onto the tether of my first mount, as our horses walked back up to the trail.

Reaching the trail, we were greeted by a laughing group. I was certain that my "take off" had been observed and was funnier to behold than to experience. Continuing up the mountain, we crossed the river and the trail became more rocky as it ascended out of the valley. The surefooted horses picked their ways carefully as the trail narrowed and became barely more than a ledge overlooking a hundred-foot drop to the stream bed below. The hikers could be seen behind when we circled around the face of the cliff.

After an hour we arrived at a large sloping grassy plain surrounded by coconut trees. Here we dismounted, tied our horses to trees, and Benwar removed the saddle from his horse. I am certain he had a self-satisfied, sadistic grin when he glanced my way as he did so. We waited for the hikers, and I was hoping I would be able to walk.

A short rest and refreshment of cool coconut milk and the white custardlike meat of a green coconut was welcome before continuing our journey. We all hiked across the grassy slopes and through swampy land amid the trunks and roots of large Tahitian chestnut trees. It seemed as though this entire side of the mountain was one large bog with tiny anastomosing rills coursing through the mud. Another large stream was crossed, but we stopped here long enough to wash off the mud and enjoy another cool drink. Looking around, we now realized that we had entered a narrowing canyon, the walls of which were becoming steeper and higher until we reached the end. We were facing an almost sheer cliff over 2,000 feet high, and we could see a small ribbon of water cascading from the top. Neither description nor photograph could possibly convey the spectacular beauty or the feeling of insignificance I felt while standing here, viewing the grandeur.

Diving in an "outer" pool, we swam under a large rock that arched over the entrance to a cavern. Inside, we could look up and see that it was formed like a chimney. We could see the blue sky far above. Two large pools, each with cathedral-like domes, were available for exploring. We could climb a large rock and jump into the deep water near where the falls were cascading into the pool; we could swim through the pelting water. Since this was the dry season, the water was not as voluminous as it certainly must be after several days of rain. During heavy rains, this would be a dangerous, roaring, frightening cave.

Our return trip was not uneventful either. While negotiating the first stream crossing, our guide suddenly stopped, held us his hand for quiet, and then shouted, "*Macho—allez.*" One of his dogs which had been accompanying us ran off and in four minutes, returned with a wild chicken in its mouth. While we watched, Benwar cut off the chicken's head, plucked and cleaned it, built a fire, and singed off the pinfeathers. The feet of the chicken were then tucked beneath his belt, and we continued our trek.

Benwar was really a man with very few words. When asked questions, he would give direct answers and that was the end. Never would he explain what he was about to do, so this day had been full of guessing, and then surprises. We were resting at the spot where the three horses were tethered, when I suddenly noticed that our guide was missing. After fifteen minutes he still had not appeared, and Mark said that he had gone into the swamp. I found him in the muddy swamp under the big trees. His hands and arms were muddy, and he was gathering flattened, nutlike objects resembling mango seeds. These, he told me, were the Tahitian chestnuts. He showed me how the husks were cut off and the shells broken to remove the nut, which was then cooked before being eaten. He was gathering these to send to his mother in Taio Hae. I offered to help fill his large gunnysack, so he showed me how to check each one for a tiny black pinhole indicating the presence of a worm; these were to be discarded. We managed to fill the sack to capacity, and I was thankful that this dirty task was completed. You can imagine my chagrin when Benwar produced three more gunnysacks, saying with a grin, "We fill."

At that point, I ran for help, but *Dulcinea*'s crew wasn't very interested and started down the trail. However, I did feel grateful to our guide, who had spent nearly a day with us, so I was happy that Brad and Russ volunteered to stay. Tiny black gnatlike flies kept biting our ankles, backs, chests, faces, and necks. They didn't hurt, but were annoying and drew a small speck of blood. Finally the four sacks were about two thirds full, and Benwar had enough!

Benwar climbed for coconuts, cut a stake, and showed us how to husk a coconut. Max had already shown us how it was done, but Benwar now wanted us to do it. Although it is more difficult than it looks, I thought we did fairly well, but our efforts struck our guide as hilarious. He laughed so hard that sweat was streaming down his face. Then, he showed us how to cut branches off a specific willowlike tree and strip off the outer covering. These were used to lash two sacks together. Two such sacks were slung over his horse behind the saddle, and the remaining two slung over one of the other animals. Benwar directed Brad to one horse as he mounted his. One horse stood with no burden to carry. "You boss," he pointed at me. "You take wild horse." I mounted and directed it with the lasso to a large rock where Russ got on behind me. Down the steep cliffside trail we went.

Once in a while the horse would slip, and I would hang on to the mane and rope with both hands. Russ would grab me, and several times we both nearly fell off the bony perch. I could feel every vertebra through my bathing suit, and knew Russ could too. We both began laughing so hard we thought we would never reach the bottom still astride the mount. Finally, Russ took off one of his tennis shoes and sat on it. Reaching the flat of the valley, we all dismounted, and the "wild horse" was released, the remaining two horses carrying the sacks of chestnuts and following the trail obediently and without direction.

Back in L'Anse Hakaui, we all felt better when we had a refreshing swim and a cup of Daniel's coffee served with ship's biscuits and butter. He told me that we were boats numbers fifty-one and fifty-two that had come into this bay during the last year. The largest number of boats at any one time has been seven. Most people don't hike up to the falls, however.

A nearly full moon lighted up the bay when two dinghies left the sandbar of Hakaui. It was good to get back "home" and to a comfortable bunk after a healthy, exercise-filled day in the sunshine and a good, clean freshwater bath in the cool, clear lagoon.

Dulcinea left the next morning, but we remained another day. The variable weather provided us with intermittent torrential downpours and hot, steamy sunshine. We hiked the beach, collected shells, and dove for lobsters for dinner. In the late afternoon, Daniel gave us a large stalk of bananas as a parting gift. We dipped it in the ocean several times to remove any spiders or insects before tying it to our mizzen shrouds to ripen.

The short trip back to Taio Hae the next morning took us an hour and a half. We faced heavy seas and rain squalls. The rain came down in such large drops that our faces stung from the impact. As we entered Taio Hae Bay, we were joined by a school of dolphins and the sunshine. Two new boats had arrived. They were from Oregon.

Most of that night Brad was kept awake by a severe itching. Inspection revealed large hivelike welts on his legs, chest, back, waist, neck, and face. Then it occurred to me. When we had been gathering those chestnuts, those biting insects were noseeums! Their bites weren't particularly annoying at the time, but now the histamine effects were showing. "Tiger balm" ointment seemed to relieve the itching. For the first time, I broke out the medicine kit; Benedryl was used. Russ and I didn't suffer until the following night. I guess every paradise has its nemesis!

I picked up my tiki from Manu and our "leave papers" from the gendarme's office. I had planned, now, to visit Hiva Oa and Fatu Hiva before sailing to Papeete. Our jib will certainly fall apart at the next hoisting, and I sure hope Nan mailed my new one to Papeete, I thought, as preparations were made for leaving. Anchor was weighed at 1720 and by 1755 we were on course 124° for Hiva Oa, the engine off, and the mizzen and Genoa pulling nicely. Our light supper of cheese and fresh-baked bread was satisfying and ample, as we had lost our "sea legs" once again.

Hiva Oa and Tahu Ata appeared with the golden dawn,

and the sun dried us out from the frequent night squalls. The pass between Hiva Oa (Dominica Ile) and Tahu Ata (Station Christiana) is named Haava or Bordelais Strait. With adverse current and winds, the engine was used to take *Pagurus II* through Vipihai (Traitor's Bay) and into Taa Huku Bay where we dropped anchor. The ninety-five-mile trip had taken eighteen hours from "anchor up" to "anchor down."

"Hey *Pagurus,* is that really you?" came a call from another boat at anchor. What a pleasant surprise! Kathy and Russ Beardsley on the *Sea Laurel* from Santa Cruz, California, our home port. They had left Santa Cruz over a year ago and had been calling at many interesting ports. They had just arrived at Hiva Oa yesterday. We joined them for a walk ashore along the beautiful big black volcanic sand beach near the village of Atuona.

Atuona is larger than Taio Hae, and the houses seemed neater, and the surrounding grounds cared for. We bought bread and tomatoes in one of the three village stores, which appeared to be well stocked. As in Nuka Hiva, however, no eggs were available.

Walking back on the road along the cliffs above Taa Huku Bay, we passed a nun with about 150 young girls ten to twelve years old. Sister explained that this was the beginning of the two-week Easter vacation and that they had been to the beach for a swim. We continued to walk through the coconut and banana trees to the beach where we had left the dinghies.

Rowing out to our boat, we stopped at a cement landing built on the shore. Large vessels tie here to load copra and unload supplies. Because of the severe surge, however, small boats would have difficulty to keep from getting smashed to pieces. It required much maneuvering to come alongside, disembark, and pull up our rubber dinghy, but the freshwater shower taken by standing under a faucet was worth the effort. Refreshed, we returned to *Pagurus II* where Brad and Russ enjoyed an orange drink and I, a Scotch.

While reflecting on the day's activities amid Gauguin's haunts, another small boat, the *Odyssey* came in and anchored nearby. Two passengers were rowed over to see us for a short visit—Rio and Charles, whom we had met the day we arrived in Taio Hae. Their story of their last week's experiences were interesting. They had departed Taio Hae on the interisland trading boat and come to Hiva Oa. After a day, they got a ride to Fatu Hiva. It must be understood that the Marquesas really is not a place for travelers. There are no hotels, and only three of the islands have airports; these are serviced only once a week by a small six-passenger plane from Papeete. Also, gendarmes are stationed on Nuka Hiva and Hiva Oa, but not on other islands. When Rio and Charles arrived at Hane Vava, Fatu Hiva, where would they stay? After talking to some of the villagers, they were given rooms at someone's house. Apparently, one night Rio awoke to find that someone had entered her room. Her screams frightened the intruder, but the next day the chief asked them to leave. How? Luckily, the *Odyssey* called for a day and brought them back to Hiva Oa. Now, they had to find places to stay here for a week until the interisland trading vessel came in. Then, they could obtain passage to Papeete, a thousand miles away.

Here, in Hiva Oa, Charles found a room at a local boy's school, since the vacation period created several vacancies. Rio found a small one-room grass hut atop ten-foot stilts, next to the sheriff's house. We all thought that she should be safe here!

The moon passed behind the peak of Mount Temeti, 3,576 feet above the sea level, moments before the eastern sky took on a reddish tinge signaling the light of dawn. I had been sitting in the cockpit observing the moon's path until it again appeared to the west of the mountain top. Bathed in the warm South Pacific breeze of the morning, I was thinking about the dinner party for Carolyn and Russ Beardsley which we were going to give that evening. It seemed like such a long time that actual plans had been made for any event. We had been swept into leisurely living from moment to moment. Planning for activities wasn't really necessary, as informality was the "rule," and yet, the mere thought of doing something special seemed exciting.

Mount Temeti became a combination of red on the steep bare lava cliffs and a green at the top where the luxuriant growth was evident. The red clouds became pink, and as the color faded, became a beautiful white against the brilliant blue sky. The gray waters upon which *Pagurus II* was suspended were also taking on a pretty blue color, reflected from the sky. Daylight had arrived.

The smell of the perking coffee woke Brad and Russ. We had pamplemousse and banana fritters for breakfast in this never-to-be-forgotten tropical atmosphere. Dishes were quickly washed, and we eagerly boarded the dinghy to explore this "land of Gauguin."

Placid waters accompanied our row ashore. This was a new paradise, with new acquaintances and new experiences. Our dinghy approached the black lava beach, behind which grew the usual coconut grove. We could see the small stream trickling from the tropical freshwater lagoon. Near the beach, beneath mature coconut trees, we could see several thatch-roofed but open huts. It looked like a movie setting of a tropical village. As the voyagers landed on the sandy beach, however, we discovered that these huts, resembling a Polynesian village, were only storage areas for various-sized boats of people of Atuona.

The path over which we walked began eastward through a coconut grove before it turned south up a gentle slope. Here it joined a road heading westward up the steep cliff toward a little lighthouse at the top. We then began a descent toward the village of Atuona at the head of Taa Huku (or Traitor's Bay). We walked past the post office and along the black lava sand beach fringed with coconut trees, enjoying the fresh ocean breeze and the warm sunshine. The main street (an unpaved dirt road) was lined with beautiful hibiscus in full bloom. We passed a small house which was marked as once belonging to Paul Gauguin, and then we walked up a steep path to the cemetery where we located the lava-rocked grave marked "Paul Gauguin, 1903," A fragrant blooming *Plumaria* shaded the resting place of this famous painter. Necklaces of strung shells adorned many graves and hung from some of the crosses.

The short hike down the path to the village was easier than the trip up, nevertheless a cold beer at one of the three stores was refreshing. We asked if any fresh beef was available, and the proprietor told us that he thought his wife had some upstairs in the freezer. The people here do not eat much meat; fish provide their major protein source. A six-pound

Gauguin's grave under a Plumularia at Hiva Oa.

chunk of beef was brought to us, and we purchased it for 255 francs (about fifty cents per pound). A gallon of wine and some vegetables completed our purchases on this day.

On the way back to our dinghy, we picked fresh limes to use in mai-tais that evening, and *Plumaria* and *Hibiscus* to add a touch of color and fragrance to *Pagurus II* for our dinner.

We cleaned the boat below, went for a swim, and took freshwater "showers" under the faucet ashore. While so engaged, about 150 young girls (ages ten to sixteen) came down from the local school. I have never seen kids anywhere enjoy the water as much. They laughed, dove off the pier, pushed one another into the water, played tag, and giggled. Their frolicking antics showed them to be as much at home in the water as on land.

Two girls appeared with guitars, and the sisters laughed as several girls tried to instruct us in the art of dancing the tamure. We were to bend our knees, move our feet, and scissor our legs at the same time—while the girls shook and wiggled from the hips down. To these young people, dance was a most natural and spontaneous means of expression of the joy of being alive. Everyone laughed, and then the girls and the sisters began singing as they left us for the long walk around the bay and back to their school.

As we were rowing back to the anchorage we saw a large trimaran coming into Taa Huku Bay. It was the *Leisure V* from San Francisco, and she was flying the "Q" flag, indicating the completion of a long passage. "Let's ask them for dinner, too!" I exclaimed. We reached their anchorage at the same time their hook was set. Although it wasn't a Polynesian greeting, our "Kaoha" and dinner invitation were gracefully accepted. Cap and Ann Smith were the owners. Rick Powell, a seventeen-year-old, and Cheresse Smoot, from Hungtington Beach, were crew.

Back aboard *Pagurus II* I tried to discern the type of meat we had purchased and decided it could be cut into steaks. These were steamed in butter, and we served tomatoes and onions for a salad, along with plenty of rice. The white tablecloth and flowers below furnished a nice setting for our dinner party. We learned that Ann Smith was a Christian Scientist and hadn't had any immunizations. I wondered what problems they would face when applying for visas the next day. (As it turned out, she had to go to a local doctor and a smallpox vaccination was administered.) Carolyn and Russ shared their experiences with us while we had coffee outside in the warm evening breeze.

The next morning we had to do something about replenishing the freshwater supply because our tanks were empty. I knew that we couldn't lay alongside the cement pier because of the tremendous surge, so *Pagurus II* was moved and anchored about twenty-five meters from the landing. Then, Russ set out in the dinghy, trailing a line from the stern and towing one end of a twenty-five-meter hose. When ashore, the line was shortened until we lay in a good position, about fifteen meters from the faucet. We now had fresh water. After filling the tanks and spare jerrycans, we scrubbed down the decks and took freshwater showers in the cockpit. Then we moved back to our anchorage.

In the afternoon we walked to the post office, where we found mail awaiting us. Returning to the boat, we saw the *Honeywind II* anchoring. Since we had met Jan and Jim McNabb and their crewman, Blake, in the Galápagos, we bid them welcome and accepted a dinner invitation for that evening.

A row to the quay for another shower to wash away the dust from the afternoon's hike was refreshing. Carolyn and Russ had just showered also. We were joined by four teachers from the girl's primary school, and one had a guitar with her. Then, Rick, from *Leisure V,* came ashore, and we were all given another tamure lesson. We thought we were learning quite well until two young Marquesan men joined in the dancing!

Brad, Russ, and I enjoyed a pleasant evening aboard *Honeywind II.* Margaritas were served in the cockpit. Then we went below to a wonderful dinner of scallops (which had been gathered and frozen while in Mexico), French-fried breadfruit, and a fruit salad of bananas, mangoes, and papayas from Hiva Oa. Jan really showed her culinary expertise when a banana-cream pie with graham cracker crust was served as desert.

Blake explained his disappointment that his wife couldn't

The sheriff's house at Hiva Oa

have joined us. She had flown from Los Angeles to Papeete and had been trying for several weeks to get on a flight to Hiva Oa, where there is an airstrip on a table-top mountain. Apparently, a weekly flight leaves Papeete for Hiva Oa and Ua Huka, but only two seats are alotted for each of these stops. Naturally, inhabitants of these islands would have priority for occupying the four spaces. In the meantime, his wife had been staying aboard the *Chantsong* in Papeete and had been calling on the radio to make nightly progress reports to Blake. Fortunately, Blake was able to obtain reservations for a flight arriving and leaving the next day, so he was going to leave the *Honeywind II.* Although these islands are not geared to handle tourists, this is really a good part of their charm, because they are relatively unspoiled by the economy tourism brings.

At 0445 the next morning, we weighed anchor, and by 0500 we were passing Hiva Oa Point on our way to Fatu Hiva. As we passed between Montane and Tahu Ata we could see the mountain peaks of Fatu Hiva, thirty miles away. The jenny and mizzen pulled us along at a good 6.5 knots, as Russ fixed banana fritters for breakfast.

A little after noon, we approached Hane Vave (Virgins Bay), Fatu Hiva. The inlet was U-shaped, about 2.5 miles long and about a quarter mile wide. The cliffs on either side became steeper, forming a 250 to 400 foot "enclosure" as we approached our anchorage. Huge monolithic giants rose up as sentinels on the port side. The anchor was dropped in five fathoms of water on a sloping, pebbly bottom. This was not the best holding substratum, but at 1235, I knew that we would have ample daylight to observe any drifting. We were facing a long, narrow valley at the head of the bay. Strong gusts of east wind made *Pagurus II* shudder as she tugged hard at her anchor line. We could hear the wind howling in the rigging. However, this was certainly the most spectacular anchorage we had ever seen. The water was deep, and the beach was steep. It was composed of stones about the size of one's fist. These rolled and clattered noisily as the ocean waves crashed over the shore. Beyond the stony beach we could see a beautiful freshwater lagoon beside which stood a white church with a red-tile roof.

When reading *Sailing Directions for the Pacific Islands,* I noted: "Violent squalls blow down the valley, and the holding ground is mediocre, vessels' anchors are liable to drag." The description did not give me confidence, so we ate a light lunch and observed the stability of the anchorage. Although the bow pointed toward shore, the ocean waves from the stern created a hobbyhorse action, and I could envision the anchor pulling out and becoming useless in deep water. Soon, since no such action seemed to occur, I judged it safe to go ashore.

The monolithic guardians appeared to watch in silence as we timed the breakers in an attempt to make a dry landing. Suddenly, with a thunderous roar, amid foaming, rushing, white water, we were heaved forward onto a sloping concrete ramp. Holding the painter of the dinghy, we waded quickly to dry land.

A path led us through the small village of about a dozen houses. Oranges, breadfruit, pamplemousse, limes, papayas, and mangoes lined the trail. *Coleus* plants grew nearly a meter high. Coconut and banana palms surrounded us. The entire narrow valley between tall mountain peaks appeared to be filled with coconut palms. It was as though we were passing between giant stone gates and into a light green valley.

The people were very friendly, and everyone shouted a "kaoha" greeting. Further up the path we met John, who explained that his father had owned "all this land." He, being the oldest son, now cares for the crops. He has four sisters and five brothers, one who was in the military service in France. Although his father was dead, his mother was still living and in Papeete. John directed us to a trail that would lead to beautiful waterfalls about three kilometers away. Brad was concerned about the anchorage, so returned to the boat. Russ and I continued on. The trail soon became an entanglement of rubble where trees had been cut, leaving huge limbs across the narrowing creek. We never did find the falls, but the view from the mountain peak was spectacular. We could look down the narrow valley and see *Pagurus II* as a tiny black dot on the water far below.

Our return hike was interrupted by an occasional dip in the fresh, flowing stream to cool off. We gathered oranges and limes until we could carry no more. The thousands of flies and mosquitoes reminded us to be wary of another apparent paradise.

In the little village we enjoyed another visit with Chief John. Would we like some pamplemousse? *"Mais oui!"* Nine pamplemousse, weighing a total of fifty pounds, was all we could carry. Fortunately, Brad had observed us and again rowed ashore!

John showed us his immaculately maintained house and his family's house, which was still used when the relatives gathered for a visit. He showed us a "very old" canoe which had belonged to his father. We were told that many people still used outriggers to travel to Hiva Oa for supplies, but in the "old days" this was much more common. John could remember his tribe engaging in cannibalism after killing an "enemy," but he said he was very young. He showed us an album with twenty-six pictures of his family. Since John was the only one in the village who knew any English, he had several letters from previously visiting yachts. These were among his prized possessions.

Another young man wanted to know if we had any *"vahinis"* aboard. He explained that the only *vahinis* here were either old and fat or very, very young. He also wanted shells or guns. I was sorry to disappoint him on both accounts!

Promising John some elementary English books about whales and the sea (which I had aboard), we took our leave. Amid white water, we loaded our fresh fruit aboard the dinghy and returned to the bobbing *Pagurus II* as the sun was sinking very low on the western horizon.

We had dinner and then a game of cards before Brad and Russ decided to turn in. I climbed into my bunk and began reading a "good" mystery. With the howling wind and the black sky, the land receded into oblivion. I decided not to spend another night here, but to leave for Papeete the next afternoon. Well, back to my mystery. Suddenly, I thought I heard a noise. Was it my imagination and thoughts about cannibals? No, There it was again! It was an eerie, high-pitched cry coming from the cliffs to the south. I looked at Brad and Russ as they lay in sound sleep. Now, I heard it again. Then, an answering cry from the northern cliffs—

"Yaaaa . . . hoooo." Now it sounded from the beach. This was too much for my active imagination, so I retrieved the crew from sleep.

They thought I was hallucinating until that high-pitched call echoed off the cliffs, only to be answered again on the other side. It sounded closer. Our imaginations now had us surrounded by savages with intent to do us harm. Russ bounded out of bed, grabbed his knife, and said, "Let 'em come. I'll get 'em." I strapped my knife to my belt. Brad picked up the flare pistol.

Peering out the ports, we could see no lights from the village, but a small dying fire was barely visible on the beach. There were no gendarmes on this island, and indeed, they only come over from Hiva Oa when there is trouble. Listening, we could hear paddles of an outrigger as they dipped into the water. I went out into the cockpit just as the canoe went gliding by, and a friendly voice yelled, "Kaoha." To say I was "startled" would be a gross understatement!

A sudden gust of gale wind caused *Pagurus II* to shudder and tug on her anchor chain, as I suddenly realized what was happening. A canoe coming from Omoa Bay, about three miles to the north, was being guided to shore on this black night by answering calls. The glowing embers of a fire acted as a beacon to guide the night navigators. Nevertheless, sleep did not arrive easily that night, and when it did, dreams of uneasy adventures accompanied it.

The next morning, Chief John gracefully accepted my gift of several books and a letter of "thanks" which I had written partly in French and partly in English.

That Sunday, at 1200, 2 March, we departed Hane Vave, Fatu Hiva, Marquesas, for Papeete, Tahiti.

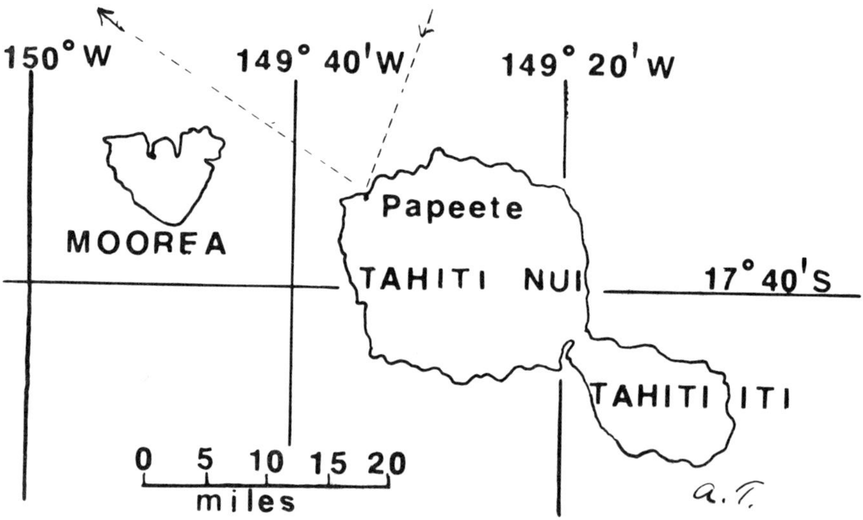
150° W
149° 40' W
149° 20' W
MOOREA
Papeete
TAHITI NUI
17° 40' S
TAHITI ITI
0 5 10 15 20
miles
a. T.

CHAPTER VIII

PAPEETE, TAHITI

Once we left Have Vave and the howling wind coming through the valley, we were in the lee of the island of Fatu Hiva, so the engine was used for two and one-half hours before the southeast trades really filled the Genoa and mizzen sails. We were on a comfortable heel and cutting through the water at a good seven knots, as we drew lots for the first three-hour watch. Russ won, followed by Brad, and then me. The first "official" watch was to start at 1500 and thus set the pattern for our passage to Papeete.

Our supper that night was simple. I fixed ham and breadfruit and made a gallon of iced lemon-tea sans the ice. The night's sail averaged 6.5 knots under clear skies on our course of 235°. Brad was on watch in the morning when I awoke to the smell of coffee and frying bacon. Russ was fixing breakfast before I was to go out on watch. We all seemed rested and in good spirits. In spite of the sea motion and the duties allowing us about five hours rest, we each slept more soundly than in the anchorage at Hane Vave, with the howling winds and vibration.

It was a beautiful morning at sea and the wind vane was holding us on a steady course, so we enjoyed a leisurely breakfast below during the first part of my watch. We were lingering over our second cup of coffee at 0920 when something happened to bring us to sudden alertness. The stillness of the morning was pierced by a loud horn. Nothing aboard

Pagurus II could have made that noise! Rushing out to the cockpit we were amazed to see a 400-foot freighter to the starboard. She could have been several days out of Panama, and the captain must have been surprised to see a small sailboat with no one at the helm. The captain of the *Clydebank* from London hailed us to ask if there was anything we needed. How comforting it was to realize that the skipper of a large vessel on a scheduled crossing would care enough to take time for such a thoughtful inquiry. I answered on the hailer, "Negative, we are all well and have everything we need. Thank you."

A loud "Cheerio" came from the monstrous vessel as it made a large circle around us and soon disappeared from sight.

That evening, our 1800 fix, using Sirius, Canopus, and Betelgeuse gave us a good position of latitude 12° 33.0′S and longitude 142° 22.0′W. During the past thirty-six hours we had made 142 miles on a true course of 238°. Our average speed was only four knots, which was disappointing, and our position was also about ten miles south of the planned course line. Since it wasn't desirable to come too near the most northwestern atoll of the Tuomoto group, I altered course for part of the day's run. One morning, I thought, we will identify Rangiroa and then change course to SSE for Papeete.

In the morning we tried our luck at fishing again. A beautiful dorado was hooked, but just as it reached the boat, the hook came off the spoon, and it was lost. This type of failure has plagued our fishing attempts many times. Right then, I decided to purchase heavier line and more stainless steel leaders when we reached Papeete. After all, we were fishing for food and not merely for sport. As I was setting another line, the aroma of Russ's baking banana bread came up from below—some consolation.

I read the four-part article on the South Pacific in the December 1974 *National Geographic.* This magazine was included in a "book trade" with the *Leisure V* when we were at Hiva Oa. From observations on our passage, it was easy to understand how the early Polynesians could have navigated by the wind, predominant wave patterns, and stars (they still do). The articles were very interesting and presented many pictures of places we had been, as well as those we had yet to see. There was a good picture of the black volcanic sand beach at Atuona, Hiva Oa, along which we had walked and stopped to cool off by body-surfing in the waves.

Three fish were hooked and lost that day, and as if to add insult to injury, Russ fixed rice and canned tuna for supper!

On 5 March, just before sighting the evening stars, we noticed a few clouds in the sky. They were to the south, west, and north, but none were in the eastsoutheast (the direction from which the wind was blowing). Near sundown, one huge cloud in the southwest was the only one illuminated. It appeared as though a fire were within its center. Even as daylight faded, the colors within the large cloud persisted as a red-orange glow. Then, in the dimming light of evening, the moisture-laden air was apparent by broad bands of reddish streaks running frum the west to the east. The weather was changing.

About 2030 I retired, but couldn't seem to sleep. Brad was on watch, and I could hear him talking to Russ. At 2045 we had a squall with heavy rain and wind, but it only lasted a few minutes. Russ came below to put on a sweatshirt and relieve Brad at 2100 (we had changed our watch pattern), but Brad remained in the cockpit until 2130. We continually checked one another during the restless, squally night. I thought of the poem:

> When wind comes before rain,
> Sun comes soon again.
> When rain comes before wind,
> Get your small sails in.

About 2230, it appeared as if another squall were coming. I was in the head when we suddenly heeled over 50°. I yelled to Brad to let out the Genoa sheet. Russ was on a course of 250°, and the screaming of the wind and water was frightening. Before I could get to the ladder or before Brad could get to the sheet, the entire clew and three gromets ripped out of the jenny. Our only good foresail!

The engine was started and the spreader lights switched

on. As the engine was set at 1,200 rpm, I wondered why most emergencies seemed to occur at night. The autopilot was set to maintain our course into the wind, as the three of us made our way to the pitching, wave-splashed foredeck. Our teamwork failed somewhere, however, because the halyard became fouled aloft. Slowly easing off on the sheet, Brad was able to get the jenny down to the deck while Russ eased off on the halyard. The large Genoa was removed from the forestay and slowly transported to the cockpit. It was then that I noticed two of the sail slides on the mizzen had let loose, so it was reefed at the second set of reef points. A seam in the mizzen had also come unsewn for about eight inches. We had some work cut out for us.

It was a mess down below. Things had even been thrown out of the quarter-berth, where we thought they were safely stowed. Books, magazines, coffee water, sleeping bags, all appeared as though they had been thrown into a huge clothes dryer. The precious chimney of one bulkhead kerosene lamp had been hit by a "flying" book, and glass splinters were everywhere, even embedded in bindings of other books.

At 2345, I took my watch and sent Brad and Russ below to try to get some rest. I could imagine how their attempts to sleep were impaired by the violent rocking with the waves. Without the full mizzen sail, *Pagurus II* would roll 25° to 30° and then remain upright until another set of waves would repeat the roll. In spite of these conditions, with the reefed mizzen, we averaged 3.8 knots during the next six hours.

It is often said that vision comes with the dawn. True, we could see, but now to get organized. The poorly recut, worn-out jib was bent on and hoisted. This provided better steerage while the main was set. Now, we would have time to remove and repair the mizzen.

I thought about finding a sailmaker in Papeete to make permanent repairs while tracing machine-sewn holes with a needle and thread on the rent seam at the mizzen's leach. What a slow, laborious task! It took most of the day. Then, I thought about the new jib which Nan had had made in Sausalito and couldn't mail to us in the Galápagos. Now I knew I would certainly have to have it sent to us in Papeete. Pan American Airlines would deliver it to me within two days.

Looking back on the events that led up to the damage to our beautiful Genoa, I could see several mistakes:

1. I should have been more acutely aware of the weather changes.

2. The Genoa sheet should be let out immediately when hit by a sudden squall.

3. At the first contact with squall winds, the course should immediately be altered towards the weather, especially when carrying full sail.

By doing these things, flapping sails may be annoying, but the terrific sudden strain of sixty-knot wind gusts wouldn't be able to put such a damaging force on the sails. This was a valuable lesson to learn.

By 1630, the mizzen had been repaired, bent on, and set. I don't remember what we had for supper that evening, but the good feeling of being once more on course at a speed of seven knots will never be forgotten.

Since I had the 1800 to 2100 watch, when it became time for sighting the stars, Brad took over the helm for twenty minutes. I took the altitudes of Sirius, Betelgeuse, Aldebaron, Capella, Pollux, Procyon, Canopus, and Achenar. Although we only needed three for a good position fix, by taking them all, we had "spares" in case one was off. Each time I neared bringing a celestial body down to the horizon with the sextant, I would call, "Ready," and then, "Mark" just at the moment when the horizon was reached. Below, Russ put down the time and recorded the altitude when I read it from the sextant. After the star altitudes had been recorded, the two who were not on watch worked out the azimuths, bearings, and distances before plotting our position on a universal plotting sheet. This night, since I was on watch, Brad and Russ worked the problem, obtained a position, measured the day's run, and course, and then recorded these facts in the course change log.

The report was brought up to me with a cup of coffee. During the last twenty-four hours we had made 120 miles on a course of 205°. With all the difficulty we had, I thought this was pretty good. Since we had been steering a course of 300° all day to compensate for the southerly set, I changed to 240° after my watch was over. This course, I reasoned, should account for the southerly set, and we should clear Rangiroa in a couple of days. Then, it would only be 181 miles to Papeete.

On March 7, during one of these violent squalls, however, the head of the jib tore loose. Brad was suffering from stomach cramps for some reason, so he took the helm while Russ and I lowered and unbent our poor, old, worn-out jib. We took it below, and after much difficulty, finally stretched the luff on the bolt wire. Using nylon webbing, I strengthened the head of the sail by hand-stitching. Then, the jib was bent on, hoisted, and we had good steerage once again.

The evening stars told us that we should reach our turning point in the morning. All during the night the sky was spectacularly ablaze with lightning, but strangely enough, though, we met only one squall at about 0415. Our speed had been 6.5 to 7 knots, so the course was changed to 195° at 0730, even though we had not seen Rangiroa. We picked up Radio Tahiti on the RDF at 74 MC and heard the 1000 news in English. This was the first English newscast we had heard since leaving Acapulco. It was interesting to learn that the exchange rate was seventy-three francs for one American dollar and seventy-four for traveler's checks. We had been getting seventy-six in the Marquesas. Truly, we were about to enter the "civilized world!"

On 9 March at 0545, the light of early dawn revealed a slight hump on the horizon just off the port bow. We were approaching the low atoll of Tetiaroa, Marlon Brando's ilsand. How fortunate, I thought, that we did not come upon this low-lying atoll in the dark of night.

Before long, the jagged, picturesque mountain peaks of Moorea appeared. By 1310, we had put out our anchor, backed down toward the quay, and were securely tied between two other boats at Papeete, Tahiti. I checked the engine log and noted that we had used the engine 41.8 hours since leaving the Galápagos. It seemed to me that traveling more than 4,000 miles and using barely forty gallons of diesel was very little drain on our petroleum resources!

We saw the *Finn,* a boat which we had seen in the Galápagos. Brunhilde and Rolf waved excitedly and came to chat. They had had engine trouble along the way, and had taken forty-five days to reach Papeete from the Galápagos. Because of their difficulty they did not go to the Marquesas.

They knew, however, we were on our way, because they had talked to Blake Chatfield when he had arrived by plane from Hiva Oa. How fast news travels, I thought.

When we left Fatu Hiva, Russ had made a card and assigned each of us to different hours of the day. The hour during which we completed docking at Papeete would determine who was to buy the first tall, cool drinks with lots of ice. I had the questionable distinction of having won! How disappointed I was when leading them to Quinn's Bar, to discover that it had been torn down. Further down Quai Bir-Hakeim, we came to the Royal Papeete Hotel. Surely, they had a bar. We each ordered a Planter's Punch, made a toast, and slowly sipped the cold, fruity drink.

It is said that magic occurs in Polynesia, and perhaps it does. Brad looked up as if he read my thoughts and said, "That looks like Dr. Chamberlain coming through the door." I looked, jumped up, and ran to greet him. Jack was a friend of long-standing. We had attended college classes together. He said that more surprises were coming and laughed. I ran outside, and on the sidewalk stood Bette, his wife, and Nan. They were both grinning. Oh, what a happy, unexpected, shocking reunion. More drinks were quickly ordered, and none could stop talking. Knowing Nan's aversion to flying, this event was absolutely unexpected.

From my letters, they had calculated that we would arrive some time during the week. They had arrived on Friday night, and now it was Sunday, and we were all here. Yes, she brought the new jib, jib snaps, shackles, sail thread, and many more much-needed items. They brought cheese and crackers from San Francisco, and a new supply of California wines. We jabbered like school kids out for a holiday. I kidded Nan and told her we had to get rid of the *vahinis* aboard!

We took hot freshwater showers and dressed for dinner. What an occasion this was. Dinner in the hotel was fair, but very expensive. Bette and Jack were going to stay for two weeks, and Nan would sail with us through the Society Islands and then back to Hawaii. We decided to keep one hotel room—mainly for the shower, of course!

The next day was exciting, as we worked as a team to locate many needed items for the boat. We also visited the market, found where we could buy ice at the Hinano Brewery, and went to the bank. For supper, we had a big salad of pineapple, avocado, bananas, and papayas aboard *Pagurus II.*

In the morning Brad and Russ found a sailmaker on a nearby boat. John Hutton used to be a sailmaker for North Sails and had his sewing machine aboard. Naturally, he was quickly invited to view the needed repairs. His estimate of forty-five to fifty dollars seemed very reasonable, and I quickly agreed to the price.

Jack helped me bend on the new jib sail Nan had brought with her. It was a beautifully made, very strong sail. However, it was about six inches too long at the foot. I could make an adjustment of the outhaul, but would have it recut when we returned home.

For dinner that evening we had a tomato, avocado and lettuce salad and filet mignon. This was the most inexpensive meat here, costing less than a dollar per pound. We set the table with a white cloth, candles, and flowers. It was beautiful, but by the time dinner was over, we had to go into the cockpit where the cooler evening was appreciated. Brad, Russ, and I demonstrated the skills we had learned by playing harmonicas.

Bette and Jack were scouting around for a good location to hold a dental convention, and in doing so, a local travel agency invited us all to a *tamaaraa* (a festival meal). When we reached Paulette Vienot's house, we were surprised to find fifty other guests, travel agents from many parts of the world. *"Iorana oe iorana,"* (Welcome, welcome to you), we were greeted as a headband was placed on each man and a fragrant garland of *tiaré* was put around the neck of each lady. The outdoor tables were set under coconut palms on a grassy, sloping area. We could see the calm green water of the lagoon washing up gently on the white sandy beach. Beyond, a ridge of coral reef separated the lagoon from the deep blue ocean.

We met "Muck" and John Kelly, two of the three owners of the Bali Hai hotels in French Polynesia. They were very friendly and outgoing, inviting us to visit their establishments when calling at the other islands. "In fact," said Muck, "you are welcome to use our hot fresh water showers at any of our hotels." It was obvious that he was familiar with the life of a small-boat sailor!

After supper, as we were sitting on mats laid out for us on the beach, a special treat had been prepared. Paulette's Tahitian dancing group came out onto the white sand to perform many of the native dances under the light of blazing torches. Amid this tropical, dreamy setting, the performers amazed us with their agility. Near the close of the spectacle, the dancers separated, and selecting partners from the "audience" completed the festivities with the rapid *tamure.* *"Iorana,"* also a farewell expression, rang in our ears as we drove back to #69 on the Quai Bir-Hakeim.

The next day we did a typical tourist thing. We boarded the Keke II, a motor launch, for the short seven-mile run to Moorea. It was a truly relaxed feeling to be a passenger and let another captain concern himself with the negotiating of the pass into Moorea. Skillfully, he brought the launch from the choppy ocean waters between large coral heads into the calm, crystal clear water of the lagoon. The first stop was the Bali Hai Hotel, at which several passengers disembarked. Here, thirteen thatch-roofed bungalows were built on stilts over the water. They blended into the background of white sandy beach, coconut palms, and towering volcanic peaks, green with lush growth. "This was such a contrast," I thought, "to other beaches of the world where high rise concrete hotels completely obliterate the background."

The *Keke II* left the dock and passed slowly through the clear placid water of the lagoon to the Aimeo Hotel, two miles away. Here, at the entrance to Baie de Cook, we could look at the end of the bay and see the jagged, green, volcanic peaks made famous by so many movies. We disembarked here and walked around the bay. Moorea has the reputation of being the most beautiful of the Society Islands, but I believe this is because it is the one most frequently visited by tourists. Later, we were to visit another island, which became our favorite.

In the late afternoon, we swam around the coral reefs to cool off before boarding the *Keke II* for the return to Papeete, a bustling city of commerce and a sharp contrast to Moorea.

On Saturday we rented a car for a drive to Tahiti Iti and all the way around Tahiti Nui. I wanted to see the Gauguin museum and the botanical gardens at Papeari. The museum was disappointing, as there were no original paintings, only reproductions. Brad, Russ, and I could, however, relate to many depictions and descriptions of Gauguin's life in the Marquesas. The botanical gardens were beautiful and specimen plants were well-labeled. It was here that I finally learned the name of the nuts we had helped gather in Hakaui. They were "Tahitian chestnuts" *(Inocarpus fagiferus).* I learned that these were not chestnuts at all, but a relative of the bean family. The solitary large seeds are boiled or roasted and throughout Polynesia are known as *ifi* or *ihi*.

A perfect day culminated in another *tamaaraa,* and when the *himaa* (ground oven) was opened, our mouths watered from the delicious aroma of the roasted pig. The evening's sunset was one of the most beautiful I have ever seen—and I didn't have my camera. The rugged outline of Moorea could be seen in the background, framed by Tahitian coconut palms and white sand. Blazing colors of the sinking sun made the clouds glow, reflecting red, orange, and yellow on the ocean's surface.

Bette and Jack were to leave in two days, and Nan was becoming apprehensive about being away from home and the long sail north to Hawaii. In the morning she decided to fly home. Naturally I was very disappointed, but the decision was made. We walked to the Pan American ticket office. "Are you apprehensive about the sailing?" I asked.

"No, it's not so much that; I just don't like to be away from home so long," was the response

"Why?"

I guess it was difficult to find a logical answer, but one was found. "Well, you plan to visit the other islands before heading north, and we have two insurance premiums due in May."

I supposed that I had been away for such a long time that this mundane problem seemed insignificant. Our oldest daughter, Linda, was married, and our youngest daughter, Nancy Ann, was at U.C. Irvine. "We could," I suggested, "send Linda two signed blank checks, and the insurance agent could instruct her as to the amount to fill in."

By this time we had reached the ticket office and were informed that seats were available on Bette and Jack's flight #816 to San Francisco. I really couldn't help Nan with her problem of indecision. Finally she said, "Well, let's go for a walk to the market, and I'll decide by 1400."

We looked in several Tahitian shops for small gifts which Nan would take home to family members and children. We saw Brad and Russ having lunch at one of the many small "Snack Shacks." Further down Rue Collette we saw Bette and Jack having their lunch. Suddenly, Nan's pensive mood changed. "I've decided what to do. I'm going to sail with you."

The decision was made and I was happy. At the market we bought avocados to use with supper. Then, we stopped at immigration, customs, and the port authority where I wrote letters assigning Nan as a crewmember aboard *Pagurus II,* an American yacht *en route* to Hawaii. Mai-tais were served to celebrate before our filet mignon dinner.

"Deux pour le airport dans Faaa," we were greeted by the taxi driver when he came to #69 Quai Bir-Hakeim.

"No, quatre," I answered, while helping Jack with the luggage. Nan and I were going to see them off. Their 1945 flight had been delayed somewhere in the South Pacific, but another would leave for Honolulu at 2200, and they could obtain reservations. With a promise to take care, I gave Jack a package containing nine months' exposed film, and with shell leis, we bid them *"Iorana."* Then, they were gone.

Nan was silent on the taxi ride back to our boat. Her commitment was complete, for I knew she would never fly home alone.

A number of details needed attention before we could leave Papeete. I had learned during our voyage that calm patience is the only solution to international official matters. All too often, we Americans are "in a hurry," which isn't at all necessary.

Russ and I had obtained our official visas long ago in San Francisco before the start of the voyage. Nan's thirty-day passport entrance would allow ample time for our visit to the windward isles. However, since Brad's last-minute decision to join us, he hadn't had time to obtain the visa in San Francisco. His passport, therefore, reflected an entrance at Nuka Hiva, so it became necessary for him to obtain a visa. In fact, his thirty-day period had passed, and I was concerned about the official problems we would encounter. As fate would have it, the chief of police was out of his office entertaining the minister, who was visiting Papeete. Two days later, he obtained the necessary stamps and signatures.

The immigration officer brought out passports, visas, and an official statement allowing us to withdraw our money (which had been deposited more than a year ago) from the Banque de L'Indochine. This also required two day's time. Monsieur Clause, the manager, could find no record of our deposits by foreign transfer! Finally, he found one for Russ, so with his 500 American dollars in hand, Russ and I left the bank.

The next day Monsieur Clause vociferously apologized and explained that he could find no record of my international order for deposit to his bank. For fifteen dollars, however, he could cable my bank and in two days he would have a confirmation. Apparently, my copy of deposit was insufficient to convince him. "Well, I will wait until we arrive back in California," I explained. I was told that I could check at the Banque de L'Indochine on the other islands we would visit, so I promised to do so, as I once more left the bank.

Borrowing $200 from Russ, I stopped at the Banque de Polynesie and converted it to francs. Then, we stopped at the duty-free shop to buy some liquor. The lady in the shop informed us that we could not purchase liquor if we were going to stop anywhere else in French Polynesia. "But our *zarpe* said Hawaii," I explained. However, it did read "via Huihuine, Raitaea, and Bora Bora." A phone call to the customs provided the authority, and the liquor would be delivered to our boat an hour before sailing.

At 1100 the agriculture inspector came to deliver our papers that would allow us to leave. Nan and Russ were stowing groceries, Brad was clearing the decks, and I was pouring over charts when the delivery boy from the duty-free shop appeared. More problems. It was decided that the sale wasn't legal because we were stopping at other islands, and if I returned to the store, my money would be refunded. The woman was visibly furious. Another phone call and the sale was again approved. By 1400, a customs official would sign the required form, and our liquor would be delivered. Well, we would wait, but at 1405, we would leave Papeete with or without our liquor. By now, the reader is probably impressed by my insatiable desire to obtain barrels of liquor at a tremendously reduced price. At 1300, the delivery boy, the lady from the liquor store, and a customs official delivered the *four* bottles of liquor I had ordered and paid for. We left Papeete, powered through the pass, and once more *Pagurus II* was in the open ocean, heading for a new adventure.

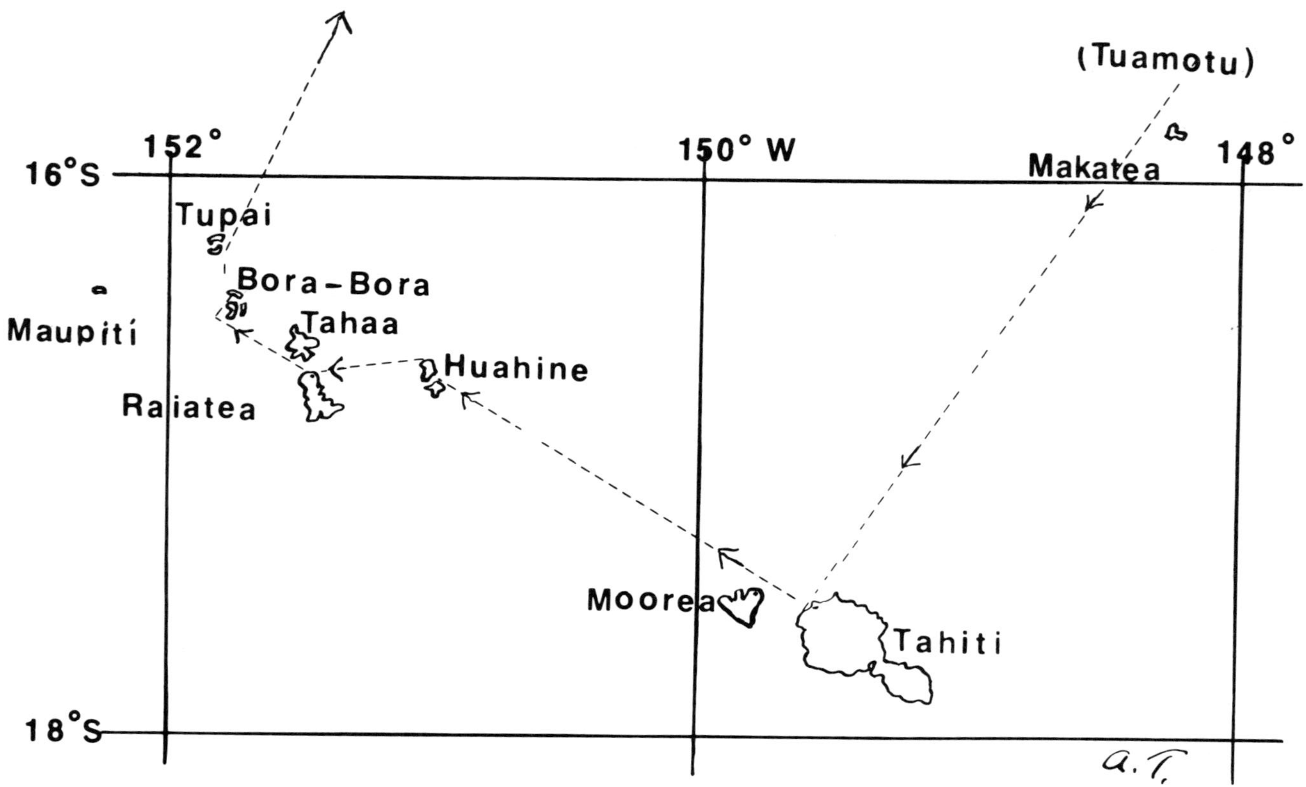

THE SOCIETIES

CHAPTER IX

THE LEEWARD ISLANDS

The winds were light, and we traveled at one knot throughout the night. By 0600 the next morning, Moorea could still be seen on the distant horizon behind us. The engine was started, and amid scattered squalls, we power-sailed on towards Huahine. I was anxious to make our landfall by daylight, because I wouldn't attempt the pass during the dark.

In the afternoon, the NW wind, which had prevailed during the day, changed to ESE and increased in strength. We could clearly see Huahine Nui to the north, and Huahine Iti to the south. A canal separating them has been spanned by a bridge, but I had planned to anchor between the islands. At 1515 the mainsail was furled, and we began to negotiate the tricky Passe Farerea. I apparently didn't find the exact pass, because Nan was reading off the depths: "fifty feet, fifty feet, thirty feet, twenty-five feet, twenty-two feet, twenty feet, fifteen feet," and then she shrieked, "twelve feet, eight feet, deep water." Wow! What a relief. I thought I was following the *Sailing Directions,* but we did miss the exact pass. We could see the rusting hulk of an interisland freighter on a sandbar, a grim reminder of someone's misjudgment, as we slowly slid by, barely twenty-five feet away. This was, by far, one of our closest calls, I thought as we entered the deep water with the easterly swells and wind.

We examined a potential anchorage on the NW side of

Baie de Maore, but I didn't like it. The wind would blow us onto shore if our anchor should slip. The SW side was more appealing. We dropped our anchor in fifteen fathoms of water near the small village of Maroe.

As we seemed to be tired, Nan fixed a pot roast, and we all had a swim before dinner. Then, coffee was enjoyed in the cockpit, as we observed a half-moon, the stars of Orion, and the Southern Cross. A cool easterly breeze was blowing scattered clouds across the star-studded sky, and we could hear voices and dogs barking on shore. Tomorrow, I thought, we will go exploring. We then said, "Goodnight" and turned in for a sound sleep. The quiet, deep waters of the bay were as calm as when we were tied up to the dock at Papeete.

Breakfast of papaya and coffee satisfied our appetites. "What shall we do today?" Brad decided to stay aboard, but he would row us ashore and then come for us whenever we hailed. I'm sure he didn't realize how long that would be . . . and neither did we. After alighting on a long pier, we walked to a church on shore and took pictures of *Pagurus II* quietly resting, her anchor line hanging straight down.

We walked through the small village toward the bridge leading to Huahine Nui and towards Fare, the principal village of the islands. We soon came to a plateau where we observed acres of growing vanilla beans. Vanilla is in the orchid family. It is a climbing vine with thick, oval, short-stemmed leaves and clinging aerial roots. The pale green flowers are sometimes hand-pollinated and produce narrow pods about six inches long. When the pods are picked green, they are "cured" by leaving them in the sun for three to five weeks. The fermentation causes vanillin to crystallize on the outside of the pod. The flower of the vanilla plant *(Vanilla fragrans)* was disappointingly nonfragrant.

Two young girls came down a driveway, and we talked to them. Sara and Valentine were their names. As we were talking, Sara's mother came down the drive and introduced herself as Lahura. She delighted in explaining the process of growing the vanilla plants. While I was attempting to translate her French, she disappeared and later returned with a handful of fragrant vanilla beans, a shopping bag, and her husband. She had decided to join us, and we would walk to Fare together. Her husband was deaf, but explained to me (in French) that she had two empty bottles. He said, "Lahura will go to Fare. Lahura will buy some wine and will return. Then we will drink wine, and we will have fun!" The bright glint in his eyes emphasized the meaning of his words.

Continuing the walk, we came to a village named Fiti on the Baie Havai, where Lahura pointed out the *école* which her daughter, Sara, attended. When we reached Fare, Lahura bid us *adieu,* and we parted.

I looked out to seaward and examined the passes through the fringing coral reefs: Passe Avapehi and Passe Avamoa. I decided, then, that Passe Avamoa would be a good approach to Ville de Fare.

We then walked to the Bali Hai and had a cold drink. "Is Mike here?" I asked. The girl ran to get him, and I introduced us. I explained that we had met Muck and Bill at Papeete. "Oh, yes," he said, "I was there too and saw you. *Iorana.*" Mike reassured me that the pass was no problem and was surprised to learn that we had gone into the "dangerous pass" to Maroe. "When you come to Fare," he explained, "be sure to use our showers." Then he showed us where they were, and as we were tired, he arranged for a taxi to take us back to Maroe for 800 francs. It was worth it.

Brad spotted us walking out onto the pier and began rowing ashore to get us. The anchorage was one of the most quiet we have had. There was absolutely no feeling of motion below. We swung, perhaps 80° on our anchor, but never felt it. After supper, we played cards until sleepy. During the night it rained, and I heard Russ get up to close the ports, but then, after the rain, it became very muggy below, so I got up and opened them again.

At 0600 the next morning, the anchor was weighed, and we followed the reef around to the pass. Nan was at the helm while Brad, Russ, and I peered over the bow for coral heads. We glided by the old rusting wreck on the north side of the pass and then were in deep water. The short trip around Huihine Nui and through Passe Avamoa required three hours, so the engine was used for the entire trip.

Anchored in two fathoms of water next to a white sandy beach fringed with coconut palms, the anchor was clearly visible on the bottom. *Pagurus II,* when viewed from the dinghy, appeared to be suspended in air. We could watch the anchor bury into the white coral sand bottom when the engine was reversed to set the hook.

That evening when dinner was about ready, a dinghy from a thirty-foot Columbia anchored nearby, came over. We met Donna, Doyle, and Scott from Seattle. Their boat was named the *Rawhide Harris.* Naturally, they were invited aboard for dinner, and we heard their story. When they left Seattle, they sailed to Hawaii, Christmas, and Fanning Islands, Bora Bora, Raitaea, Huahine, and arrived in Papeete three months ago. They liked it so well here, in Huahine, that they had just returned. When they leave here, they planned to go to Fiji, Samoa, and then to Australia. Doyle's son, Scott, was a junior in high school, so plans had been made for him to return home in September. They were thoroughly enjoying their voyage and apparently had no time schedule.

It was cooler that night, and we all slept well. In fact, I didn't wake until 0630, which was late for me. Brad did the dishes from our night's dinner, and I dried them. Then we had coffee and went in to the Bali Hai for a shower. Later, in a little gift shop, we met a young couple (Fran and Dave Yacheson) from Saskatchewan, Canada. They were on a tour of the Leeward Islands, and were to become good friends. Nan, being friendly and gregarious, asked them aboard that afternoon for cocktails. That day, we swam, walked, showered, and swam again. Brad and Russ walked to the village and consumed apricot nectar, ice cream, and cookies. Wow! The young people can eat all the time. When I saw a large watermelon at the store, my enthusiasm got the better of my judgment, and I decided to have it for dinner. I thought it was seventy francs, but the Chinese storekeeper said, "No, six cent." She meant, "six hundred."

"No, tres cher," was my response, so it was not purchased.

With the dinghy, we picked up Fran and Dave on shore and had a good visit aboard *Pagurus II.* Then they invited us for dinner, so we rowed them back to the beach and said we would meet them later. It was fun to dress for dinner. Nan wore a long blue gown and Brad, Russ, and I wore shoes, long pants, and flowered shirts. We then managed to row ashore and land on the beach without mishap. Dinner was good, but we really weren't used to eating so much. Pea soup, marinated fish salad, pepper steak, potatoes, assorted cheese, coffee, and cake made up the fare. At 2130 we bid Fran and Dave goodnight as we set out for a walk along a little path that led through the coconut grove and into the village. Next, we walked to the place on the beach where we had left our dinghy. The exercise helped to ease the strain of such full stomachs.

On another day, Brad, Nan, and I decided to go to Maeva. Russ, on this occasion, wanted to remain aboard, write letters, and relax in solitude. Therefore, three of us were put ashore in the dinghy, and, as Russ rowed back to the boat, we walked the path along the shore to Fare. When we climbed aboard *le truck,* an obese woman offered us beer, which we gracefully refused. Loaded with passengers, produce, live chickens, and people, *le truck* bounced along the dirt road of Huahine Nui, making frequent stops to let off various passengers. The driver indicated a stop where we could get off and explained that we could view the many historical interests as we continued on the road to Maeva.

We hiked and examined the rock fortifications and old graves of the ancient chiefs of Polynesia. Huahine was, at one time, a sacred island of the Polynesians, and many stone and coral slabs marked individual worshipping places called *maraes.* It was hot, so we walked out on the fringing reef for a swim. Returning to Maeva, a passing bus was hailed, and I asked if he was returning to Fare. *"Oui, dans demi-heure,"* was the response. Thereupon, he took us to a watermelon patch where *le truck* was loaded with melons, leaving barely enough room for us. After a delicious slice of melon, the driver, two workers, Brad, Nan, and I boarded for the return to Fare. We learned that an interisland ship was arriving in the morning and the fruit would be transported to Papeete where it would fetch high prices. We could believe this because we thought the prices in Papeete's market were very high.

Something seemed strange when we reached the beach opposite which *Pagurus II* was anchored. We noticed that she had moved. Russ was at the helm, and the engine was run-

The remains of an old Maori temple, "marae" is made of coral slabs in Maiva, Huihine.

ning. Brad and I swam out to the boat and assisted Russ in his efforts to reanchor *Pagurus II.*

We finally reset the hook in three fathoms of water on the coral-sand bottom and made several dives to be assured of an adequate set. When we rowed in to get Nan, we found her sound asleep under a coconut tree.

Then, Russ and I went to the Bali Hai for a shower, and here he treated me to a Polynesian punch with rum, pineapple juice, and lime. He commented, "Now, I think I understand why you say that you need a drink after coming through the passes."

Later, as Russ recounted what had happened, we all listened intently. He was busy writing letters below when he felt that the motion of the boat had changed. Looking out the port, he was surprised to discover that the anchor had dragged, and he was about two-thirds the distance from the shore to the reef. This happened about a half-hour before we arrived on the scene. Reanchoring a boat by one's self is no easy chore, and he was happy to see us meandering along the beach. How fortunate, I thought, that Russ decided to stay aboard.

On a visit to the Banque de L'Indochine on the next day, the banker said, "No word yet on your international money transfer, but you should check in Raiatea." Well, it seems as though the time was right for leaving Fare, so I'll check with the crew and see what they would like to do, I thought. They wanted to go to Hawaii!

"Now, wait a minute," I explained. "We have come this far, and to leave without seeing Raiatea and Bora Bora would be regrettable."

By 0930 the next morning, we were negotiating the pass leaving Fare, and by 1400, *Pagurus II* entered Passe, Uturoa, once more entering placid waters of a tropical lagoon protected by a fringing reef. The channel was well marked. At 1500 we dropped anchor in fifteen fathoms of water and rowed a stern line ashore to make it fast to the base of a coconut palm. The waters of the cove were calm, and we could see a large building being constructed near the point. This was to be the Uturoa Yacht Club. The breeze came from the starboard stern quarter, blowing down the companionway hatch, making it very comfortable below.

Another small boat, from Sweden, was anchored to our starboard, and not far away was a boat from Mexico. We didn't have an opportunity to visit with these people, as they were busy preparing to leave in the morning. They did, however, show us the location of a freshwater, open-air shower.

My charts showed our location to be next to an island between the fringing reef and the main island of Raiatea. However, a causeway had been built from Raiatea, and this island was being developed for recreation. Between the causeway and our anchorage, an olympic-sized pool had been constructed of coral and concrete. One side was left open to the waters of the lagoon. Students from the local schools held swim meets in this large, beautiful pool.

"Let's check to see if the Banque de L'Indochine has found our money," I said as we finished breakfast the next morning. On the three-mile walk into the town of Uturoa we found the office of the gendarmes, where we officially checked in. Then, passing the post office, we stopped to buy stamps and mail letters. The bank was modern, and the manager pleasant. He phoned Monseur Clause in Papeete. Still no luck! We were requested to check again when we reached Honolulu. However, the bank had received and held mail for us, which was a most pleasant surprise. Letters from home would highlight any day in the life of a sailor. Right then, I decided, however, to wait until we reached home, and my bank could ferret out the problem.

Borrowing another $100 from Russ, I crossed the street to the Banque de Polynesie, where it was converted to 7,350 francs. This certainly was not a good rate of exchange, but the Banque de Polynesie had very friendly tellers, and I wasn't willing to give the business to the Banque de L'Indochine.

We met Scott, Doyle, and Donna from the *Rawhide Harris* in the shopping area. They said that their anchorage was in a little cove not far from us and near the airport. Being less protected, however, they had more wind than we. Also, we were apparently following the same route as our commercial traveler friends from Canada. Dave and Fran were in a small grocery store where we stopped for supplies. Arrangements were made for them to come to the boat for dinner that evening. Then, the following evening, we were invited to their Bali Hai bungalow for cocktails after which we would all go out for dinner. The experience of anchoring in strange islands and yet finding friends and acquaintances was most pleasant. Wherever we visited, however, we had our snug "home" with us.

Raiatea is the most significant of the Leeward Islands, and Uturoa is the commercial center and administrative capital. From the west, the Maori migrations first landed on Raiatea, and it was from here that the Polynesians spread to other Pacific islands. A large *marae* is also located in a small village of Opoa, thirty kilometers from Uturoa. Historians tell us that this ancient Polynesian temple site was visited by pilgrims arriving in double outrigger canoes from as far as Hawaii.

Mount Tapioi overlooks Uturoa, the nearby islands, and many lagoons. Ships call at the deepwater port of Uturoa, tying up at the docks. We could see several large ships of the French navy anchored near the main channel.

It seems as though, while cruising, much time is spent in obtaining supplies. We located a marine shop that carried Perko supplies, so I purchased a pull-electric switch to replace one that had broken. Then, we bought some heavy plastic five-gallon containers and had them filled with a total of sixty-three liters of diesel to supplant our deck load. Each of us carried one container as we walked the two miles along the hibiscus-lined road back to the anchorage.

Sharing the late afternoon with Fran and Dave at the Bali Hai was fun. We could see Passe Uturoa through which we had entered two days before. We dove off the porch to snorkel on the coral reef and swim among the many beautifully colored tropical fish. Leaving the water after our exercise, we climbed the steps leading to a small landing where salt water was rinsed off under a shower. Four more steps took us to the bungalow's deck where Fran had arranged tall cool drinks and snacks. I thought of San Francisco, the Golden Gate bridge, and the busy freeways. They all seemed to exist in another world, obliterated by surrounding beauty of the natural setting before us.

About 2230, after dinner, we attempted to get a taxi to convey us the seven kilometers back to *Pagurus II*. "Impossible," we were told. A large wedding was being held, and every available taxi had been commandeered for the evening.

The Bali Hai's manager came to our rescue when he learned of our plight. The distance itself was not disturbing, as we had walked in, but the hour was late and we were tired. The manager's offer of a key to an orange Peugeot, therefore, was gratefully accepted, so Dave and Fran drove us back to the Uturoa Yacht Club.

We threaded our way through a beach party in progress before reaching our dinghy and rowing out to the boat. Through the remainder of the night and early morning hours we could hear the guitars and singing of Tahitian songs, as the revelers enjoyed themselves. This Good Friday in Uturoa, Raiatea, had been a most satisfying day.

By Sunday morning, all was in readiness to leave for Bora Bora. I guess we will have to call this island "Captain's Choice," because if I were to utilize democratic rule, the crew would have us leave for Hawaii today, I thought. Although our lives had been leisurely and most pleasant, island paradises are not home, and the desire for family, familiar surroundings, and even routine, can pull strongly at one's heartstrings. Perhaps we're not nomads after all, but need the security provided by the accumulation of life's efforts. Philosophically, it was an interesting consideration, because there was absolutely nothing materialistically that we needed, and we were all healthy human specimens. Brad, Russ, and I had been away longer than Nan, but our roots would always be in California, and I knew that some rainy day in my study at home, the fire would be burning cheerfully as the recollection of these thoughts created mixed emotions. It is happening now, as these words are put down in writing. Are we never satisfied? Yes, I believe we are, but the human, with capabilities of memory and association, can look to the future, plan more adventures, and then return to snug surroundings.

At 0800 on 30 March, more than 100 fathoms of anchor line were stowed, and *Pagurus II* powered away from the secure cove of the developing Uturoa Yacht Club. We headed for the channel markers indicating navigable waters between giant coral heads of the large reef separating the islands of Raiatea and Tahaa. The island of Tahaa had many interesting anchorages I would liked to have explored, but I didn't want to push my good-natured crew too far. We negotiated Passe Papai on the western side of Tahaa and headed for the visible, but distant, jagged peaks of Bora Bora.

At 1615 we set the anchor (after three attempts) in very quiet water and only a stone's throw from Hotel Bora Bora on Matira Point. The white sandy beach was dazzling in the late afternoon sun. The jagged ancient volcanic peaks reflected on the placid lagoon waters, giving the impression of being suspended in air. We could see the bottom through crystal clear water.

Fran and Dave had arrived the day before, and we had arranged to take them out to dinner on this occasion. Brad and I rowed in to the landing and, at the hotel's desk, inquired if they would honor my American Express Gold Car. "*Mais oui*" came the answer. We made dinner reservations and returned for Nan and Russ. Cocktails were enjoyed at "Anua." This was the name above the doorway to Fran and Dave's over-the-water bungalow at Bora Bora. Here we met Nancy and Bill Megowan, from Encina, California. They were fellow commercial travelers with the young Canadians. We learned that Nancy and Bill had owned an Islander sloop and maintained a buoy at Avalon, Catalina, although the boat's home berth was at Marina Del Rey. Naturally, we discussed boating adventures and had an enjoyable visit. As a sailing man, however, I was saddened to learn that they now owned a forty-three-foot cruiser with twin Chrysler engines!

Our festive dinner held better conversation than food, and when I presented my American Express Gold Card, it was refused. The girl at the bar was most definite and stated that the cashier providing me with the earlier information was "crazy." The cashier's office was closed and the situation was rather embarrassing, but I was determined. Russ offered to pay, and Dave wanted to put the bill on his room, and Bill said he would take care of the dinner expenses—my dinner! "No," I thanked them. "May I please speak to the manager?" I asked.

"But sir, it is late, and the manager is at home," she responded.

Not quite ready to give up, I asked to see anyone who was in charge of the hotel. A young woman soon appeared. "Hello, I'm the manager, and my name is Kathy. May I help you?"

Once again I explained the situation while the "gang"

went out for a stroll on the paths beneath the luxurious tropical growth.

"Are you guests of the hotel?" asked Kathy in a pleasant, but efficient-sounding voice.

"No," and I explained for the third time.

"Are you going to be here in the morning?"

"Yes, for several days," I answered. I then gave her permission to take possession of my prized American Express Gold Card until morning, at which time the cashier would be on duty and a satisfactory arrangement could be made.

In the morning, the cashier, who was thoroughly acquainted with such transactions, had the necessary form requiring my signature to obligate the debt, returned my card, and I took my leave without an apology.

Brad helped me work off my upset condition by snorkeling with me over the reef. In the clear, calm waters of the lagoon, we exposed four rolls of film, using my underwater Nikonos. Unaware of the passing time, we were surprised to surface after a dive and see an outrigger approaching us. Nancy, Bill, Fran, and Dave were calling on *Pagurus II* for afternoon cocktails.

Everyone must have enjoyed the late afternoon, because it was 1915 before the visitors shoved their outrigger off into the darkness and headed toward the hotel. In the morning they were to board a plane. It was hard to realize that in two day's time they would be in Los Angeles, and that Fran and Dave would be home in Saskatchewan in three.

The next day we wrote letters, changed the engine oil, added oil to the gear box, and I checked out the autopilot. We scrubbed the bottom, glued a broken spoke on the wheel of the helm, and repaired the automatic bilge pump by installing a new impeller. We were nearly ready for our 2,000-mile passage to Hawaii.

In the midafternoon we rented a car to drive into the village of Vaitape for last minute supplies. Mount Paia loomed over the village, its jagged peak of 661 meters pointing into the blue sky. I was relaxed, and the crew was happy. Tomorrow we would leave for our northward passage to Hawaii.

CHAPTER X

THE NORTHERLY VOYAGE

The radio is a marvelous aid to the navigator. All throughout the South Pacific we had been able to receive WWV to get an accurate time signal. The night before our departure from Bora Bora, I checked my, less-than-a-year old, wonderful crystal digital watch. This trusty timepiece had never been out of its case, and as it truthfully reports the seconds, it had been a faithful aid on our voyage. On this night, however, it was 10 minutes, 00 seconds slow. I read the directions and changed the batteries, after which it completely refused to register. "Well," I decided, "we would have to check WWV daily and record the accuracy of one of the other watches to use for navigation."

On 2 April, at 0740, the anchor was weighed, and *Pagurus II* powered through Baie Povai, Bora Bora. Passe Teavanui was negotiated at 0810, the sails set, and a course of 280° was taken. It was evident to all that our once acquired "sea legs" had been lost. The winds carried us along at six knots through the seemingly rough seas.

By noon we passed the atoll of Tubai to our port and changed course for 005°. The passage from Bora Bora to Hilo, Hawaii is not known to be an easy one, and I wanted to make as much easterly as the winds would allow. I had hoped to be able to intercept a line from the Marquesas to Hawaii about the time we reached the equator. Then, I predicted more favorable conditions for a smooth passage for the remainder of

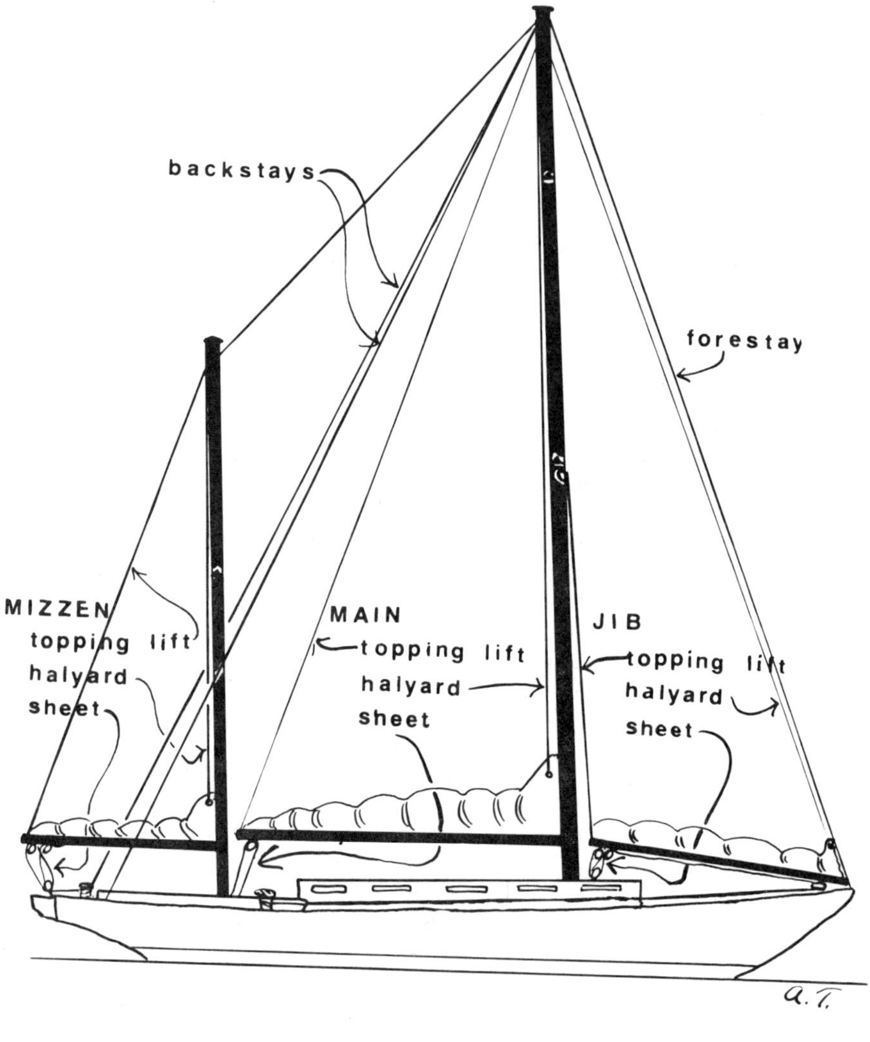
backstays
forestay
MIZZEN
topping lift
halyard
sheet
MAIN
topping lift
halyard
sheet
JIB
topping lift
halyard
sheet
a.T.

the trip. The charts clearly showed that we would experience a westerly set for the first days because of the south equatorial current's westward flow.

On the second day at sea, rain squalls and strong winds accompanied us. We were, however, averaging seven knots, and we were able to steer as much as 10° and 15°. Nan was troubled with mal de mer and taking very little food in spite of antinausea tablets. I managed to convince her of the necessity to take fluids, so she had some tomato soup and a cracker. Also, she spent most of the time in her bunk.

We had to select a new set of morning stars for the celestial position fix because the Southern Cross was too low in the horizon. Antares and Shaula were good ones to use.

On 4 April we had a good steady night's sail with smooth seas, and in the morning Nan was feeling better. She ate breakfast of soft-boiled eggs, bread and butter, and some tea. Later in the day we all had a regular soap-down shower during a heavy squall.

For the next few days we made steadier course and speed at night than during the day. Sometimes during daylight *Pagurus II* glided slowly through calm, but undulating, waters at barely one knot. The boredom was broken one day when I ran the engine for four hours.

Once, when Russ was on watch, he woke me about 2230 and said he thought he saw the shadow of islands on the horizon. We all got up and peered into the darkness with an uneasy feeling of not being certain. Even though we concluded our apparition was caused by horizontal clouds, I knew that Vostok, Flint, and Caroline Islands were in the vicinity, and since we hadn't been able to get good celestial fixes because of the cloud cover, our apprehensions were not without cause. None of us slept much that night.

We powered for the next day, and the steady travel gave Brad an opportunity to bake. The aroma of yeast bread and cinnamon rolls soon permeated the boat and came wafting into the cockpit in a most tantalizing way.

That night a good fix was obtained using Sirius, Canopus, and Rigel. We were only five miles from our D.R. (Dead Reckoning) position. This was remarkable, as it had been two days since a suitable position fix had been recorded in the log.

I decided to use the engine for the next day and night. Then, we would add some of the deck load to refill the tanks. At 0145 in the morning, a change in the sound of the engine startled me to alertness. Russ, who was on watch, explained that the engine suddenly revved-up to 3,000 rpm and power was lost at the same time, so he cut back on the throttle. The engine was shut off as Brad and I took off the companionway ladder. Then, we cleared out the heavily stowed after-locker so I could remove an after-bulkhead to get to the engine. Sailboat engines are notoriously difficult to reach, and this one was no exception. However, once this valuable stowage area was cleared, I could, at least reach most parts of the "iron monster."

When I checked the fluid level in the gear box, I was amazed to find it completely dry. Transmission fluid was added, the gear put in forward, and the clutch released, thus turning the shaft from the force of the water passing over the screw. I now noticed that the seal of the gear box was leaking transmission fluid where the copper line came out, leading to the gauge in the after-control and switch house by the helm. Tightening the hex nut at the junction stopped the leakage. More attention will have to be given these connections, I thought. The gentle NNE breeze was still too light for sailing, so we continued to power.

The one-quarter waning moon and the myriads of stars were beautiful. It seemed to be the "season" for falling stars. On two occasions these were exceptionally bright and left long trails that lasted for several seconds, appearing as comets. I wondered if any other person on earth had viewed these dying celestial bodies.

Since I had the 0300 to 0600 watch, I didn't sleep. About 0545 Nan came out to enjoy a beautiful sunrise with me before taking the helm.

After stopping the engine at noon, we were lucky if we had traveled five miles by 1800. Bowditch *(American Practical Navigator)* says that the equatorial doldrums are very narrow during February and March and may even be nonexistent. I certainly hoped this condition would last into April this year. We were 7° 30.0′ south and had experienced doldrums for the last three days.

On 7 April at 1800, Sirius, Canopus, and Rigel provided us with a perfect "textbook" position fix: latitude 07° 32.0'S and longitude 148° 34.0'W. We were right on track.

The next evening a little breeze came up, and we were soon doing a steady four knots. By morning our speed had nearly doubled, and we could hold a true course of 010°. After the past few days of stagnation, the feel of the heel and speed was exhilarating. Although we were still 420 miles south of the equator, I was hoping that we had entered the NE trades. To add to our joy, we were joined by a school of porpoises. Then, to our surprise, we noticed a fishing boat going SE about five miles off our port side.

Although the days were warm, the breeze was cooler than in the Societies, and a sweatshirt felt good at night. The sheet-blocks and tackle, the booms, and part of the masts and sails appeared as shadows on the white foam flowing along the leeward side, as the bow sliced through the water. The morning sun was warm, as we were pushed northward by the NE winds. Nan liked to sit on the coach roof soaking up the warmth from the sun.

In the evening, after sighting the stars and plotting our fix, we enjoyed chatting in the cockpit. Several times Brad made a bucket of popcorn to add to these pleasant, reflective moments. It was surprising to realize how much light the stars produced. Even on the nights when there was no moon, the few scattered clouds could be seen, and when a bird flew overhead, its size and form could be discerned. We could see the horizon for 360° around us.

One morning the engine was started at 0600 to charge the batteries. We took advantage of the power to lower the mainsail and resew four sail slides and reinforce an area where the grommet was tearing from the sail. With the engine at 1950 rpm and the jib and mizzen pulling, our speed was from 8.2 to 8.5 knots, so we weren't losing any time while making repairs.

About midnight on the ninth day out of Bora Bora, we would cross the equator, but we decided to initiate the one poor pollywog during the daylight hours. I prepared a card of certification for the crossing. Brad made a cake, and I fixed Polynesian curried chicken and rice for our dinner. Russ harvested alfalfa sprouts to mix with garbanzo and kidney beans for a salad. Then Brad, Russ, and I put on paper hats, and I used my pareu for King Neptune's cape. When the pollywog couldn't answer questions with proper answers (as was *always* the case) a nod from King Neptune resulted in a dab of colored frosting on her face from one of the attendants, and then a paper triangle would be placed in the frosting. It was truly a gentle ceremony for a gentle pollywog. The shellback was then given a drink and a piece of chocolate cake to conclude the ceremony. We crossed the equator at longitude 149° 10.0′W.

The sky became covered with cirrus, altostratus, and cumulus clouds, and by evening, long red streaks running north and south filled the sky. The winds became lighter, and course was changed to 340° true. For the next four days we weren't able to obtain a star sight because of the overcast.

The winds picked up and so did our speed. With jib and mizzen alone, the speed varied from six to nine knots. The heavy, nearly abeam seas often splashed into the cockpit, wetting the helmsman. Although the water didn't feel cold, we were soon constantly wet and even developed a saltwater rash from the rubbing clothing. After two days, things below began to get wet, and even the carpet became sloshy.

By the third day of almost steady gale, Nan wasn't even leaving her bunk. At one time a huge wave broke over the cockpit, sending a torrent of water below and even the sleeping bags got soaked. The chill of the northern air was a sharp contrast to the South Pacific warmth to which we had become accustomed. The rains became very cold.

Now, sitting in my study with a pleasant, cheery fire, I must also recall the unpleasant three-hour watches (especially at night) when cold rains came down, there was a sudden lifting of *Pagurus II,* a loud slap as a wave caught her sides, and hundreds of gallons of water completely obliterated the sight of the boat's cabin. I remember the sound of water draining the cockpit and anxious voices from below calling in unison, "Is everything all right?"

I can still envision small breaks in the clouds, through which I could perceive a glimpse of the new moon or bright Venus casting a soft glow on the ocean waves. I remember

Nan's feeling of depression. "It's too rough. . . . I just want to get to Hilo. . . . How long will this weather last? Oh. . . I need a Dramamine."

Two more days of the same clouds, overcast, heavy winds, choppy seas, and rain. We were still under jib and reefed mizzen. On 17 April we finally got a position fix placing us less than 600 miles from Hilo. Although this fact was encouraging, everything we did seemed like a very difficult chore. It was difficult just to climb out of the bunk, get dressed, and go up to relieve the previous watch.

It was at this time that we were joined by ten pilot whales that seemed to "enjoy" the weather. Much to our concern, they would circle the boat, approach it from the windward side, and then pass under the keel to again appear on the lee side. This was the third time on our entire voyage that we had observed pilot whales in this "dolphin-like" behavior. Below, we could hear their high-pitched sounds. We were relieved when these large mammals left us.

On 20 April we could get six Hawaiian radio stations, but even this and the encouragement of 350 miles to go did not lift our spirits as much as I had hoped, because the winds and seas remained the same. During one three-hour night watch I counted five storms. Since the day after we passed the equator, we had been thrown about, tossed off balance, and wet. Sighting the stars was especially difficult. And then, working out the problem presented more difficulties, sometimes requiring as much as an hour and a half. The unpredictable lurching would change the focal distance between the eyes and the nautical tables in the book, so it became difficult to focus on the required figures. Naturally, arithmetical errors were made more frequently, so parts of many problems had to be reworked.

In spite of our uncomfortable progress, we managed to cook, but Nan didn't eat. One night I made a tamale pie. The next night, Russ made a white sauce to which he added fine herbs and two cans of Campbells garden vegetable soup and a little cheese. He put this over cooked macaroni shells, and we had our dinner in bowls. Truly, it was good!

We picked up a weather broadcast from a radio station at Hilo. "Gusty NE trades thirty to thirty-five knots with seas running four to twelve feet. Small craft advisories." Didn't we know it!

By the evening of 21 April, we had 200 miles to go. The voyage was beginning to seem like a never-ending nightmare. It is really too bad, I thought, that Nan was not on the passage to the Marquesas, rather than this one.

I was on watch and alone with my thoughts at 1830, contemplating boat construction and things that could be done to make such a passage more comfortable. At 1900 there was a sharp noise, and a wire hit me across the shoulders. Turning into weather I yelled for Brad and Russ to come up on deck. The main backstay had parted at the fitting on the masthead, so both starboard and port backstays were down. The engine was started, and the jib and mizzen furled until we could survey the problem and find a solution. It appeared a sail was needed to steady us, so the fore and aft motion (hobby-horsing) wouldn't cause such a strain on the main mast. Since the mizzenmast was self-supporting, I raised the mizzen sail and this resulted in a steadying action. This sail alone was pushing us along at five knots, and all appeared to be well when another sharp report sounded. The head of the mizzen sail tore loose from the halyard. Again, I yelled for Brad and Russ, as I began clawing at canvas to furl the mizzen. There wasn't much to do to repair any damage while it was dark, so the engine was again used through the night. I really began to think out various solutions now. With 180 miles to go against wind and ocean swells, we could only make three or four knots at very best. It was now doubtful whether we would have enough fuel. In turn, we each tried to climb the mizzenmast to retrieve the halyard without success. When Brad, Russ, and I were all on deck again at 0600, I attached the bosun's chair to the mizzen topping lift. Russ and Brad, then, hauled me up the swaying and wildly arcing mast, and I succeeded in catching the halyard. With wire, needle, and sail thread, I put in a new ring, and the mizzen was once again set.

The backstay was another problem, and quite serious, as we couldn't set the jib or the main because the strain would be placed on the two small backstays of the mizzen, and I was sure they wouldn't last. Using the halyard of the mainsail,

Brad and I hoisted the backstay on the windward (starboard) side. This starboard backstay was also the radio antenna, so I thought this might be to our advantage. Once this was done, the jib was hoisted, only to be furled again when we observed an approaching squall.

It was at this point that I decided to call Honolulu Coast Guard Station. I had planned to advise them of our position and condition. I would inform them that we were certainly in no immediate danger, but if they didn't hear from us within three days, a search could be started. My attempts to contact them failed, however, and after an hour, the radio suddenly went out. Something in the transmitter kept blowing out the fuse. Well, I thought, we could receive and hear Honolulu announcing the foul weather, but we could not transmit.

We were on our own. At 1800 the engine was turned off, as we were down to about twenty gallons of diesel. When the seas calmed, we would again hoist the jib for the last 150 miles. For more than 1,000 miles we had had wearisome seas and bad weather. That evening I set the sea anchor, and we hove-to all night. We really rested comfortably, and the three-hour watch shifts were merely lookouts. In fact, if it hadn't been for the groaning and creaking of the main mast, I believe I could have slept soundly.

In the morning we set the jib and averaged about four knots. The seas were still heavy, and just as I thought everything had stabilized, there was a loud crash as a huge breaking wave tore off the companionway dodger. Then, there was another sharp "crack" and Russ yelled, "Al!" I thought the mast had splintered. Dragging my wet, cold, bruised, and tired body up the companionway ladder, I slid open the hatch to view the damage I knew awaited me. Brad followed, and Nan came close behind. She wasn't to be left below! It wasn't as severe as I had envisioned. The jury-rigged backstay on the starboard had parted, taking down the main halyard and the main topping lift with it. Russ started the engine, and I immediately instructed him to shut if off—there would be danger of the trailing lines entangling in the screw. We hauled in and coiled the shrouds and lines. Then I noticed how the masts were whipping as we pitched, so we furled the jib as quickly as possible. Without the jib, I could take the mizzen topping lift and attach it to the stern chain plate. This did succeed in reducing the whipping motion of the main mast.

That evening we could see the beacon light at Cape Kumukahi, but by morning we were no closer. This windward side of the island offered no possible shelter, and we were fighting the southward current along the southeastern shore of Hawaii. I couldn't imagine how we could possibly make Hilo, which was the customs and check-in station for boats coming from foreign shores. "Well," I decided, "our safety comes first; I'll forget about customs and think about what other solutions are available."

Looking again at the charts, I noted that the currents flow southwesterly along the southeastern shore of Hawaii, and then northerly along its western shore. There must be safe anchorages on this western or lee shore. I decided to try to make it to Kealakekua Bay. From here, I could walk to Napoopoo and contact the officials—informing them that we were home. If they could deliver fuel to me, I could power to Honolulu, but that mast must be secured first.

While Russ manned the helm, Nan and Brad hoisted me up the main mast on a bosun's chair attached to the jib halyard. At the upper spreaders, fifty feet above the deck, I clutched the mast for a few minutes to catch my breath. The boat looked like a toy, and as I peered down, I could see the water at one time, then the boat, and again, water on the other side, as the rolling and pitching of the boat caused the mast to swing and pitch. Suddenly a realization came that the situation wasn't so bad. All I had to do was to hang on with one hand and work with the other. It must be done. I must say, however, it was slow work, but the swaying and lurching no longer made me dizzy.

Reaching one hand down under the seat, I found the line securing a block and tackle, the three heavy lines trailing down to the deck. I secured one block to the mast by passing two loops of one-inch nylon line around the mast above the upper spreaders. The lower block was fastened to the chain plate of the starboard mizzen shroud. Another block and tackle was likewise secured on the port side.

Just as this had been completed, I heard Nan call my

name and looked down to see her standing in the companionway with a camera. This moment had been recorded.

Returning to deck, I tightened the lines, and the mast was secure enough to fly the mainsail if we had wanted. However, by this time I had decided that Hilo was out of the question. We had already passed South Cape and had only twenty-six miles to go. The current had really helped us on this southward track.

We brought sleeping bags, mats, and towels up from below to try to dry them off in the sun and breeze. It was much warmer in the lee of the island. If only our fuel holds out, I thought. Nan even stayed above, curling up in the sunshine on the cabin top. I often think that we must have presented a strange sight to any observer on shore. Surely we were a seagoing laundry boat!

We were really tired. I'm sure it wasn't only the physical strain of hard work and no sleep—the mental strain was really wearing. I had had a headache for two days. I was also quite concerned about coming "home" without checking into customs, as required by law, but I felt that the emergency justified my action.

At 1740 on 24 April (after twenty-two days), the anchor was dropped in Kealakekua Bay near Napoopoo. Each of us commented that the surrounding stillness was almost overwhelming. We fixed clam chowder and cornbread before falling into sound, dreamless sleep.

Reality returned with the dawn, but the serious problems and worry seemed to have left us. Brad and I rowed ashore and obtained a ride up to the highway where we walked to Napoopo. Nobody sold diesel here, so I called Standard Oil at Kailua. I was informed that they wouldn't be able to deliver diesel to Kealakekua, but if I could come to the pier at Kailua, they might be able to deliver it there; permission would have to be obtained from a superior source because it would be Saturday—I had forgotten what day of the week it would be! In the meantime, if I could come to the plant in Kailua, he would give me enough fuel in my jerry cans to be sure we could power up. A later call confirmed an appointment to have the Standard Oil truck at Kailua pier at noon on Saturday.

Several attempts to contact Mr. Charles Auld, the customs official at Hilo, failed. I then tried to get a taxi without success, and a building inspector overheard us discussing the problem. He kindly took us to Kailua, where we obtained sixteen gallons of diesel, and then he took us all the way back to the beach where we had left the dinghy.

We were back aboard by noon, transferred the diesel to the tanks, and weighed anchor at 1300. The short trip to Kailua was comfortable and relaxing. We enjoyed looking at various buildings on the rugged shore, and dropped anchor at 1600 amid several small boats.

It was 1645 when I finally reached Mr. Auld by phone from the King Kamahamaha Hotel. After I explained the emergency, he granted my request to leave for Honolulu immediately after refueling the next day. He asked for names and birthdates of all aboard and then asked if we had ample provisions aboard. I replied, "Yes, but we would really like some fresh fruit and vegetables." Mr. Auld kindly gave me permission to purchase these supplies.

For dinner on that occasion we had lettuce and tomato salad with blue cheese dressing, and steak. Oh yes, we also had V.O. with Seven-up and lots of ice.

Reanchoring at the fuel pier the next morning was really tricky business. The surge was too much to allow us to safely tie to the dock, so we put out a bow anchor and then rowed a line to the pier. Then, by shortening the stern line, we came to lie within five feet of the solid cement pier. I was happy that the Standard Oil truck arrived early (at 1115), and we took 103 gallons of diesel.

By 1320 we cleared Hawaii and set a course of 316° true for Honolulu, only 150 miles away. Using jib, mizzen, and engine, our passage was uneventful. What a thrill it was to see Diamond Head and approach Honolulu the next day. Not knowing where to go to properly check in with the officials, *Pagurus II* was taken into the commercial ship channel and tied to the Coast Guard pier so I could seek information. Really, my thoughts were: "Oh, to get to a shipyard!"

As you will soon see, my crew had other thoughts at that time.

CHAPTER XI

A HAWAIIAN ADVENTURE

After being disabled at sea and replenishing fuel at Kailua, it was a relief to enter Honolulu Harbor at 1300 on Sunday, 27 April. Unable to find a "customs dock," I tied up to the Coast Guard pier and went into the office to ask for the procedure of reentry. A young coastguardsman woke up another who was sleeping on a couch. Neither knew what should be done. Finally, one called customs at the airport. The customs official here did not know the procedure either, but he said that I must stay where I was and could not move the boat. Naturally, the coastguardsman said that this would not be allowed because I was tied to the U.S. Coast Guard pier which must be kept clear in case of an emergency. When this information was relayed to the airport customs official, his helpful voice came back, "Well, that is a problem between you and the state and doesn't concern me at all." Then, I told him that I would call back later when I found a berth somewhere.

Walking to the harbor pilot's office, I found all the doors locked. I could see him standing on the dock beside his boat, but couldn't attract his attention, so I returned to Pier 4. At the head of Pier 4 was an office of the military police. Thinking that they might have some information or, at least, know whom to contact, I explained my problem to a friendly marine. He made a few phone calls, and soon I was talking to Bill Townsend, the harbormaster of Honolulu Harbor. I

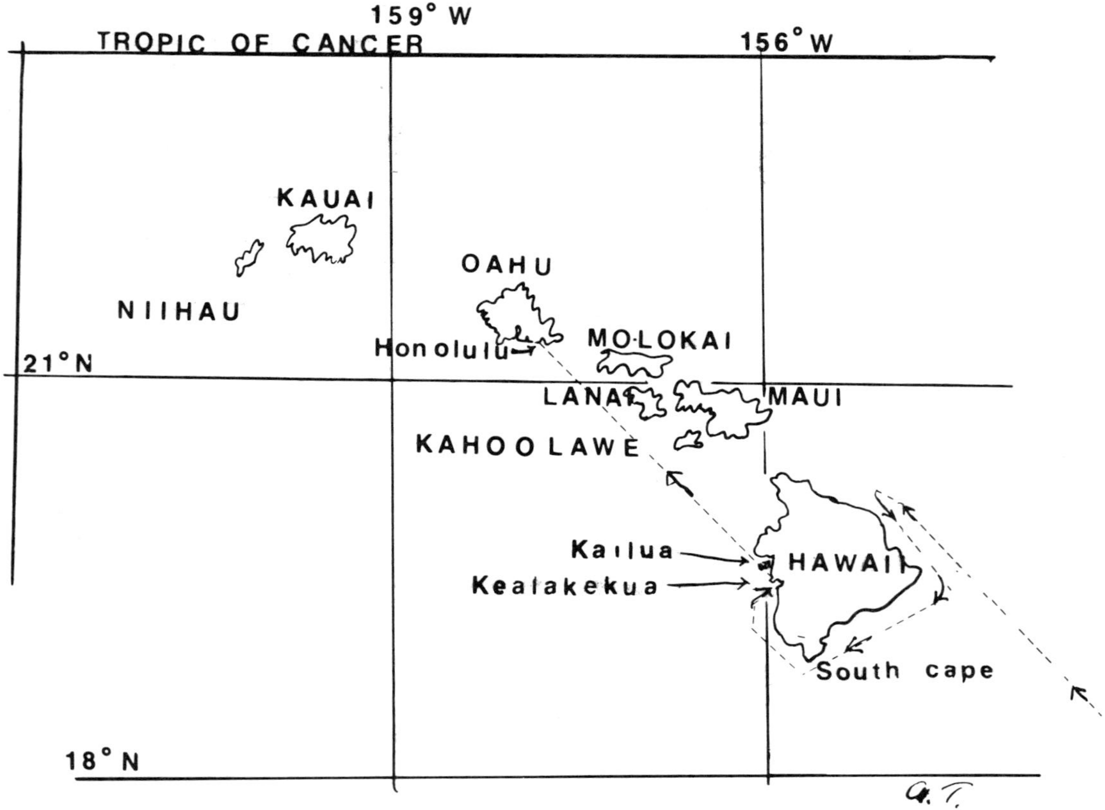
159° W
TROPIC OF CANCER
156° W
KAUAI
OAHU
NIIHAU
Honolulu
MOLOKAI
21° N
LANAI
MAUI
KAHOOLAWE
Kailua
Kealakekua
HAWAII
South cape
18° N
A.T.

thought he was most pleasant and friendly—especially when he was given a problem on a Sunday afternoon while he was entertaining guests at home. He explained that I had entered the commercial harbor, but he would see what he could do for us. My call was returned in five minutes, and Mr. Townsend had assigned us to Pier 39-C and said that a harbor policeman would be waiting there to assist us.

When *Pagurus II* reached Pier 39-C, a freighter was docked and unloading cement. After two passes, the policeman (on shore) informed us that he had been in contact with Mr. Townsend and, because of the dusty condition at that location, he thought we would be more comfortable at the foot of the pier. We tied up alongside an eight-foot wall behind a Korean fishing vessel. The permanently installed fender of the wharf protruded far enough to push against our shrouds, so we rigged a "camel" from a wooden pallet and floated this on the water between the pier and the boat, to keep us further away from the dock.

Another call to airport customs and a most friendly lady official greeted: "Aloha, welcome home." Behind me, Nan kept prompting, "Ask her if we can go to a hotel." Much to my surprise, we were given permission, but cautioned not to remove anything from the boat. We should check in with Marine Customs at 0800 the next morning. After thanking her, another phone call resulted in reservations at the Ilikai.

The Koreans were very interested in our "tiny" boat that had been sailing for nearly a year, but our communication was quite limited. They kept pointing to our jury-rigged backstays and talking among themselves. Their navigator was interested in our method of celestial navigation, so I showed him the HO-211 worksheets and gave him one. They were truly overjoyed when I presented them with a carton of American cigarettes.

At the time, we left in a taxi, I knew *Pagurus II* would be safe with our new Korean friends nearby. Fifteen minutes later, we drew stares from people in the lobby of the Ilikai as we registered. Our connecting rooms were luxurious. Two showers were immediately occupied while I called room service and ordered crackers, dip, and drinks to be sent up. The room appeared to be swaying as if we were still at sea.

Dinner than night at the "Top of the I" was served in elegant style, and we enjoyed the complete relaxation far above the lights of Honolulu. Even though the dining room seemed to sway, the trials and problems faced at sea seemed to vanish. As we talked about some of our experiences, recollection of many details became hazy.

At eight o'clock the next morning, Russ and I were at the customs office near the base of Aloha Tower. The official was very disturbed that we had left our boat, and he wanted reassurance that nothing had been removed. His attitude was suspicious as though he thought we were criminals. Having had enough of the rude encounter, I explained in a very positive way that we were Americans returning home after an extended voyage, and that officials had been informed every step of the way, and that we had been given permission to leave the boat, having informed the officials where we would be staying.

Then, I asked him what procedure he would have recommended. Apparently most boats just come in and tie up at the yacht club before contacting him the next day!

We met the agriculture inspector at Pier 39-C and all was in order. The customs officer who was also to have met us never did arrive. At any rate, we were now "officially" back home.

We located the harbor master's office in the Aloha Tower and met Mr. Townsend. He was an extremely busy but patient and gracious man. He explained that we could stay there another day until we found a berth and would only be charged a nominal port-entry fee. He also made phone calls to inquire about boat yards where repairs could be made, and we decided that the Ala Wai would be most convenient.

At the Hawaiian Yacht Club we were told that we could obtain two-weeks' visitor's privileges. However, it was suggested that these arrangements should be made after our repairs so that our allotment wouldn't be used up while in the yard.

A rather long list of needed repairs was made and presented to Mr. Peter M.A. Arapoff, manager, Ala Wai Marine, Ltd. He agreed to do the work and showed us the shops, explaining how the bill would be computed on the required work. An estimate couldn't be given, but he "guessed" it would be around $2000. We arranged for docking space and a haul-out date the following week.

Pagurus II was brought to the Ala Wai Yacht Basin. Brad and Russ would remain to oversee the repairs, and I would fly home with Nan. I had business requiring attention, and also wanted to arrange conferences with several students.

During the week at home, Brad called twice, to report progress on the work. On Monday 5 May, at 1630, *Pagurus II* came out on the marine railway. Since the hour was late, Brad and Russ scrubbed the bottom. As often occurs, yards are slow in accomplishing needed repairs, so Brad and Russ (at Pete's suggestion), spent the next few days sanding and putting on two coats of Vinelast blue bottom paint, sanding and varnishing the doghouse, and installing a new hand bilge pump.

The yard removed the masts and suggested new rigging for the main mast, because weak points were detected in the heat-swaged fittings. Since this would be a considerable added expense, Brad deferred decision until I returned the following week.

On Thursday 8 May, Linda (our eldest daughter) came back with me for a week's vacation. We retained rooms at the Ilikai, as the boat was too dirty for comfort, and the luxury of beds and hot showers was worth the expense.

Two estimates were given for a new rigging. One was for a reduction in the size of the stainless steel wire rope and one was for "replacement as is." However, the difference in cost was only a couple of hundred dollars, and experience had shown that strength of supporting structures is most important when the seas begin to show their force. I asked if the chafing sheaths could be installed when the rigging was made. Assured that this was no problem, I ordered the new rigging, as well as a new dodger for the companionway hatch.

We removed, coiled, and labeled the shrouds, forestay, and backstays of the main mast. Again, I reminded them not to forget the chafing tubes. Larry, the rigger, said that he thought there was only one backstay and had calculated his estimate on one. I reminded him that our contract was "replace as is," but would certainly pay for the additional length of stainless steel wire rope if he had, indeed, misfigured. The inefficiency of some workers today seems to plague me, I thought.

The next week when the new rigging was brought to the yard, no chafing gear had been assembled on it. Why can't I get what I request, I wondered. I would have to talk to them about this.

I had a birthday, and festivities were planned. Linda, Brad, and Russ treated me to a good dinner, after which we attended several discotheques. I was teased about my "Hawaiian five-0" and shall always remember that occasion with fondest memories of a wonderful, thoughtful crew. My fiftieth could only have been more perfect if Nan had been with us.

We were scheduled to go back in the water on Wednesday, 14 May. Pete told us that he wanted to do it near the end of the day when the tide was high. At 1530, Linda, Brad, Russ, and I went to the yard. After waiting an hour, Pete asked if we would mind waiting until the following day which we did. Then *Pagurus II* was returned to the water. She looked so strange without her masts, as she bobbed gently beside the pier.

For a celebration, I took the crew to the Polynesian Culture Center for the evening. The spectacular dancing represented cultures from many South Pacific Islands. When we watched the Tahitian *tamure* being danced with the proper vigor, we laughed, recalling our feeble efforts.

Saturday morning we took Linda to the airport and bid her an "Aloha" farewell before returning to work. First on the agenda was to clean up the mess. Sanding dust covered everything. Paint cans, brushes, sandpaper, equipment, buckets, and tools were everywhere. Nowhere else can a boat get in such dirty disorder as in a yard. Everything was taken to the cockpit while vacuuming, washing, and dusting was performed below. We wanted to move back aboard, but until some semblance of order was achieved, this would be impossible.

On Monday, an "electrical technician," came to inspect and repair our autopilot. He said he was familiar with the unit, but later in the afternoon, he left, having borrowed the manual to study the schematics.

Tuesday, as the rigging was brought aboard, I asked the worker about the chafing gear. He said he didn't know anything about it but would inquire. In the afternoon, Brad went to the airport to check "General Delivery" for mail, but Russ was with me when a most disturbing event occurred.

At 1630, Pete came to the dock and said, "I was told there's some problem with the rigging and it is something to do with the shroud covers." I briefly explained the situation.

"I don't want no hassle. Here's what I'm going to do," he said. "I won't take any more———and I'm not going to hassle. I'm going to stop all work on *Pagurus* and make up my bill, and you are going to pay me and then when you settle with the rigger, you can pay him, because I don't have anything to do with it."

I was truly astonished and invited him aboard where we could discuss the problem quietly, but he refused. I also invited him for a drink, and he refused. In fact, he even got more furious. He said, "I am not going to take any more———and you are going to pay up, and I don't care if you stay here three days or three months."

I explained that my contract was with him and not with subcontractors, and that he had told me that his yard was capable of effecting all needed repairs. I asked him if a customer would request something to be done and then had to accept anything that was turned out.

"Essentially, yes," was his answer. Then he tried to tell me that the chafing gear was no good and only served to increase the windage, and he wouldn't have any of that———on his boat. He further said, "I talked to Park and he said that the yard is not responsible in this matter, so I am making out your bill, and you will pay it up." (Park was one of the yard's most responsible workers.)

At this point, I explained that I thought his attitude was unfair and that we should discuss possible solutions, as I could not possibly pay him now, because I had only three checks left. One was for him, one for the Yacht Club, and one was for me, since I was almost out of cash. (At that instant, I realized that I would have to have Nan send me a few more checks!)

He repeated, "I'm not going to take any more hassle, and that is all there is to it. I'm making your bill now, and you can talk to Park and Larry [the rigger] to settle that."

Less than a minute after Pete had left, a neighboring boater came over to ask, "What was that all about?" I was very brief in my answer, but the entire "explosion" was uncalled for and embarrassing. I am sure all the boaters around thought we were being "called down" for having done something very bad.

At 1700, Park came aboard. On the deck, he examined the turnbuckles and said, "It is not unseamanlike to put a toggle above these, and then the rigging could be shortened . . . or the toggle could be put aloft." We discussed other alternatives to the solution of this problem which I still felt was not of my making. I further explained my feelings about the maltreatment of a customer by the manager. Apparently everyone knew he was this way, but this awareness did not negate my feelings nor my embarrassment over the entire incident. Park was truly a gentleman, and said he would talk to the rigger and think about it overnight. He would return in the morning with other possible solutions.

When Brad returned about 1730, we had much to recount regarding the afternoon's unpleasant encounter.

On Wednesday morning (21 May) Park returned about 0900. I felt his embarrassment over the situation. During the night I had decided not to go along with the toggles and to have the yard install the rigging as it was made, deliver my dodger, complete the autopilot repair (if possible), and to get out of the yard. I knew that there was really no compromising with the situation, and my personality revolted at such disharmony.

Park offered to supply us with four-year's supply of white split-vinyl covering material for the main shrouds. He said that he realized the lengths were only six feet and, therefore, shorter than those I had on the boat. I agreed that this was fine, and further explained that, at this point, I wanted all the work on the boat completed as expediently and expertly as the yard could do it so I could leave. He thought that we could have our masts stepped (reinstalled) within a couple of days.

On the morning of 22 May, Pete came to say that they would step our masts. We moved up to position, and after two hours waiting, the mizzen was put in place and supported by two shrouds before everyone went to lunch. In the afternoon, the main was stepped. Pete was really a master at handling the crane with the "dangling" sixty-foot mast. Once in position, three men worked until closing time, but the shrouds on the main didn't seem to fit.

The next morning Brad and I went to the post office and the bank. When we returned at noon, the rigging was still not installed. Larry, the rigger, was aboard and couldn't understand why the forestay was too short. Although fearing to create another disturbance, I did mention several times that the upper main port shroud appeared to be too long. Larry and Park explained that it wasn't tightened yet. Inspection, however, showed me that it was tightened as far as the turnbuckle would allow. Everyone went to lunch. Park returned to paint some spots on the mast which weren't done, put split-vinyl coverings on the shrouds, and then he went to the yard to begin painting the booms which I had had reglued.

I stayed below reading most of the afternoon because the whole mess was becoming repulsive to me. At 1440 I heard Larry yell, "This port shroud still has a label saying 'forestay.' " At 1450, Brad hoisted Larry up to the top of the mast to exchange the short "forestay" with the too long "port shroud." Once this was done, the stays and shrouds were in correct positions and fit beautifully. Of course, the wooden donut, which had been cut and glued, was now on the forestay. "Don't worry," comforted Park, "I will fix that with no trouble at all."

Larry and Park left about 1700. All shrouds and stays were in place, but had yet to be tuned.

Another problem now developed. When the new track was put on the boom, a wood bedding piece was not used. The cars would not slide. Also, when the clamps were used to tighten the tracks on the after-ends of the booms, the track became bent, so even if there had been enough space below the track, the cars would not slide on the track. Several cars were filed to make them slide, but not satisfactorily; also, the filing definitely weakened them. "Don't worry," we were told, "this can be remedied with no trouble at all."

What strange and disturbing experiences we were having. We had been in Honolulu less than a month and had met (it seemed) more difficulties over which we had no control than in our entire nine months' cruise.

On Saturday, Brad, Russ, and I tuned the rigging, carefully tightening the turnbuckles on alternate sides, occasionally stopping to walk up on the Ala Moana bridge to sight the position of the masts. The new dodger was installed; it had been well constructed, and it fit beautifully. Brad and Russ changed the engine oil while I installed the new boot tops. I had sewn these canvas fittings when I had returned home. They fit securely over the bases of the masts and serve to prevent water from coming below. After several coats of blue paint, they would become waterproofed. Brad and Russ then "moused" all the turnbuckles. This was done by threading wire through a hole in each turnbuckle and twisting the two ends together, thus preventing vibration from loosening the fitting.

The autopilot was repaired and seemed to work well. Dave found, in the pilot house control, a wire that had coroded and was not making proper contact. He explained that it was a "power lead to the limit switch."

By Sunday morning *Pagurus II* was beginning to look almost seaworthy. The booms were not yet installed, as they were still in the shop waiting for the tracks to be corrected. Brad and Russ went to church, but I remained aboard to cut, glue, and refit the teak cockpit grating, allowing for the new bilge pump pad which had been installed.

When Brad and Russ returned, we left for an afternoon and evening with friends. We had been invited to join them at the Garden Room of the Honolulu Hilton. An afternoon's entertainment by a New Orleans dixieland band and surprising guest performers provided an enjoyable atmosphere. All problems vanished in the happy tone of the afternoon.

On a thirteenth floor apartment balcony at the Ilikai, we had our dinner. A "fire-pot" was shared as we cooked Chinese chard, onions, scallops, beef, clams, shrimp, and red snapper. When noodles were finally added to the steaming broth, we had a deliciously flavored soup.

By the time we returned to *Pagurus II,* still berthed at the dock behind Ala Wai Marine Ltd., it was after midnight. Happily tired and sleepy, we said goodnight and climbed into our bunks for what was to be a sleepless night.

Awakening from the deep first hour's sleep, I heard Russ say, "What are you doing?" Sitting up and trying to focus my eyes, I saw Brad sleeping and Russ sitting up in his bunk. Then, I saw Russ standing at the navigation table. Suddenly, I realized that it wasn't Russ who was standing. The "Russ" in the bunk and I both jumped out of bed as the intruder sprang up the companionway ladder. Still dazed and groggy, we followed him across the Ala Wai boatyard to the steps leading to Ala Moana Boulevard. Here, our uninvited "guest" stopped briefly to put on some shoes, ran up the stairs, and loped in a trotting fashion across the bridge.

Halfway over the bridge, Russ and I suddenly stopped. We were clothed only in our undershorts. Quickly returning to the boat, we woke Brad, and the two decided they would give chase. However, if an accomplice were near, I thought, it would not be prudent to leave the boat. At 0215 I called the police and told them that I was now concerned about Brad and Russ chasing someone through Ala Moana Park. In ten minutes, seven police cars were in the park and another came to the boat for a description of the invader: "medium build, about five feet, six or seven inches, wearing shorts and a white T-shirt." This is all I could tell them. A call came over the radio. The police had found Brad and Russ; I was relieved.

"What was stolen?" I was asked. We hadn't even had time to look, so began a search. The contents of Brad's wallet were scattered all over in the galley, and we found his wallet in the quarter-berth. My wallet had been in a drawer of the navigation table and was intact. Russ couldn't find his pants or wallet. Apparently, the fellow was leaning over Russ when we awoke. Naturally, a long search turned up nothing. Fortunately, Russ did not have much money in his wallet, but all his identification cards and driver's licence would have to be replaced.

Later that morning, I left a note for Pete:

Dear Pete:

We are moving over to the Hawaii Yacht Club where I hope it will be safer, as we had a thief come aboard in the early morning hours while we were asleep. We weren't able to apprehend him, but he got away with one wallet and a pair of trousers. The police were contacted and were out to see us.

I didn't know about the problem with the boom tracks until Saturday afternoon. When I examined them I was amazed to see that no pads had been placed under the thinner stainless steel track. As you are aware, all throughout our repairs I have said that I want everything done in the best possible way. Our expenses have been high, and I have never complained except when the caliber of the work was in question.

Also the cars have been filed down so much that they are considerably weakened. Please devise a corrective procedure which will not involve weakening these cars.

We had hoped that we could have had our work completed so that we would be able to take a short sail, return to retune the rigging, and be able to leave by next weekend, as we are anxious to get home.

I will come over to see you Tuesday morning, but am leaving this note so you won't be concerned when you find *Pagurus* gone.

Sincerely,
Al T.

We moved to the Hawaii Yacht Club, where we dropped anchor and rowed a line to the shore. We could merely step from the stern to a small concrete pier nearby. It was peaceful here, and we could look over the sloping green grass leading up to the clubhouse. We could also look across the Ala Wai channel and see several boats anchored near the shore of Ala Moana Park. As the location was away from the street, I felt more secure here. It was strange to realize that, after traveling to so many foreign ports, we had returned home to have such a disagreeable event as to have a thief come aboard while we were sleeping. I had never before even considered such an occurrence.

On Tuesday morning at 0830, Russ, Brad, and I were at the yard. I really didn't know what Pete's reception would be, but he was very amiable and friendly. He said that he understood the problem and would have it fixed because he knew we were anxious to leave. He also explained that he would have the cars for the track filled in because they never should have been filed so thin. Nothing was said about the cost of the labor, however, and this was coming to be a concern of mine. When I was paying by the hour, to have a job done, the added cost of having it redone could be considerable. I decided, however, to approach him on this matter after the work had been completed.

Pete asked if there was a track pad on the booms before, and I replied, "No, because we had a heavy brass track, and it stood higher, but the replacement with a stainless steel track made the difference."

"Don't worry; we'll take care of it," he assured me.

"We'll stop by later this afternoon to get an approximate completion time," I suggested.

"Good; good," was his parting comment.

We had many errands to do on this day. Brad went to the airport post office in hopes of finding a letter from Lynn waiting for him at General Delivery. Russ and I washed the carpets, stowed the old rigging, added water to the battery cells, added transmission fluid, stowed two baskets of spare parts, and scrubbed down the galley. Then, we even had time to do three loads of laundry before Brad returned to sand the doghouse sides and apply another coat of varnish.

An Erickson sloop came in and anchored next to us. She had been anchored at Lanai, and the captain had an accident in which he broke both legs. A Coast Guard rescue helicopter flew him to Honolulu, where he was hospitalized. His two crewmen were novices and didn't know how to sail the boat, so his poor wife had to find someone with the knowledge to bring the sloop back to Oahu. They had been on their way to the South Pacific for a voyage of several years duration. Now, she didn't know whether to sell the boat or pay a crew to return it to Long Beach, California. Thinking about such a catastrophe made me thankful for my competent crew and our healthy, accident-free voyage.

In the evening, Mike, a worker from Ala Wai Marine Ltd., came by to notify us that the booms were ready. When we went to the yard at 0830 on Wednesday, Pete was jubilant, frivolous, and amiable. He informed me that he didn't add the hours of Tuesday's work, and I told him that I thought this was only fair. He asked if there was anything else. I told him about the two shrouds incorrectly installed and about the time required to correct this, so he deducted four more hours. In spite of these adjustments, labor cost was $2,682.00, and the total bill was $5,090.04!

Pete let us use his truck to convey the booms to the yacht club and said that the truck was ours to use whenever we wanted. We took advantage of his offer by going to the grocery store and provisioning the boat for our return passage.

With booms in place, sails bent and covered, *Pagurus II* once again looked ready. We called home that evening to appraise everyone of our anticipated departure date: 2 June.

By Thursday evening, we were ready for company we had invited for dinner. At 1530, as I was preparing Polynesian curried chicken ahead of time, a telegram arrived. Karl Wolf (an M.D. friend who had prepared our medical kit many months before) was going to join us on the passage to California. He would arrive this evening at 2120 on United Flight #129. Could we meet him? Now, what to do?

Brad and Russ both offered to go out to the airport to meet Karl, but they would miss dinner, which was to be served about 2030, and I needed help in serving the meal. When I phoned information at United, they told me that a message could be delivered to a passenger. Good. "Karl Wolf. Take taxi to Hawaii Yacht Club. We are berthed at visitor's dock. *Al.*"

The Hawaiian sunset was beautiful, and we served snow cheese, crackers, almonds, and cocktails in the cockpit. About 2030 I thickened the sauce for the chicken and added cashews, onions, bell peppers, pineapple, mandarin oranges, and a little orange juice. Then I steamed Chinese vegetables, and dinner was served below on a white table cloth with

Plumaria blossoms for beauty and fragrance; Rosé wine for libation. Date-nut bread and creamed cheese was ample for desert.

About 2215 our guests left, and I was disappointed that they had not met Karl. He arrived at 2230, carrying his gear and walking. "What happened?" I asked. He said that when he asked the cabbie how much it would cost to go to the Hawaii Yacht Club, the answer was, "Seven-fifty." Karl said, "I only have six-fifty, so take me as far as that will go." Karl walked the remaining distance!

On 31 May, Brad and I went to the airport to check with "Weather Central." We examined the weather charts for the past week and determined a fairly stable condition. The high pressure area between Hawaii and California appeared to extend as far as 30° N latitude.

Later in the afternoon I called Nan to wish her happy birthday, and tell her that we would definitely leave Monday.

Monday morning we went to the grocery store to purchase perishables. I was surprised to find that these items came to eighty dollars. We needed a taxi to help transport them back to the boat. By 1030 we were ready to leave, but the secretary of the yacht club wouldn't be in until 1300. There was nothing to do but wait. However, she came in early, so I retrieved my hundred dollar deposit, turned in our keys, and we weighed anchor to power to the fuel dock. We took only twenty-six gallons of diesel, 300 pounds of ice, and we filled the water tanks.

As we were ready to cast off the lines, Margie Bower, from the yacht club, presented us with a Hawaii Yacht Club burgee and a loaf of home-made bread. This was a wonderful parting gesture which wouldn't be forgotten.

Exactly at noon, *Pagurus II* left the fuel dock, swung her bow into the Ala Wai channel, and headed out to sea. Halfway out the channel, a beautiful sailboat came up fast from our stern and along our port side. It was Park from Ala Wai Marine Ltd. He wished us a good trip and "Aloha" before coming about to return to the harbor. This was, indeed, a thoughtful gesture and pleasant send-off.

We power-sailed to Barber's Point and then set a course

for 322° true until we reached Kaena Point at 1900. Then, the course was set for 012° true, and we logged eight to nine knots through the night.

CHAPTER XII

THE SUNSET OF A VOYAGE

We had been using the engine until passing Kaena Point at 1900. Now, heading 012° true, the winds were favorable, and we logged seven knots. The quiet swishing of passing water was a relief when the engine was turned off, but it seemed as though the bilge pump was running too frequently. Removing the companionway ladder and the after-bulkhead, I inspected the shaft and noticed that the stuffing box was leaking. In fact, this area, where the shaft passes through the hull, was admitting a regular stream of water. Further inspection revealed that the mounting bolts holding the engine tightly to its mounting were loose. So, I thought, this is why I sensed an unusual vibration when we were running the engine. I tightened the bolts and repacked the stuffing box, which stopped the leak.

I then started and ran the engine again to check the effectiveness of my repairs, but was disappointed to see that vibration continued, and the stuffing box again began to leak. Apparently the motor mount was loose. Repacking stopped the leak when I shut off the engine. "Well," I thought, "we could return to Honolulu and to the yard, or we could continue." Since I really did not like to think of facing more yard work with delay and possible unpleasantries, and since I didn't plan to use the engine until coming into the harbor at Santa Cruz, I decided to continue. After all, sailing vessels shouldn't have to depend on engines!

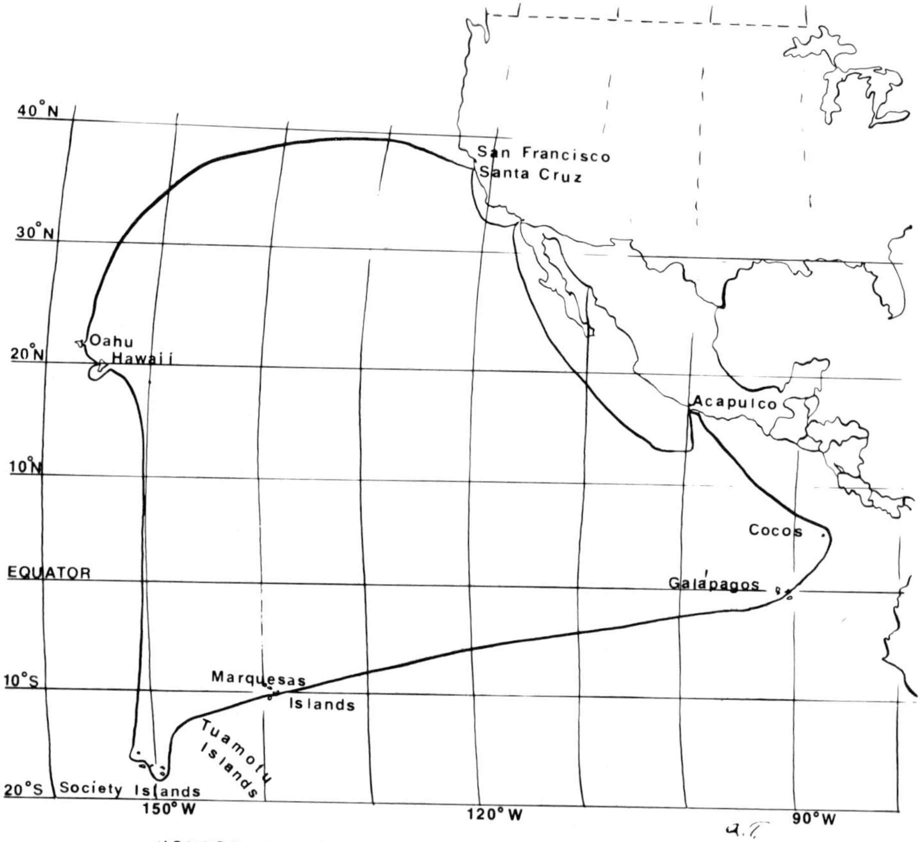

VOYAGE OF THE PAGURUS

Outside, the wind was blowing thirty to thirty-five knots, but the seas were not high. Occasionally *Pagurus II* would plough her nose into a wave that was higher than others, and those below could hear green water running over the decks.

With the dawn came the warming sun in the almost cloudless sky. Brad, Russ, and I seemed to quickly regain our sea legs, but Karl was experiencing motion sickness; he even lost his peanut butter sandwich. Alternating periods of sleep, fresh air, and sunshine seemed to help him overcome the malady.

My sailing plan was to proceed north until we noticed a wind shift and then to pick up a more easterly course, thus circumnavigating the Pacific high pressure area of fair skies and no winds. This high pressure area increases in size and its northward location in the summer months so that sailing vessels often have to travel north as far as the latitude of Washington before finding the NW winds which will carry them south to San Francisco. According to the data observed at Weather Central in Honolulu, the high pressure area was developing slowly, so I hoped we wouldn't have to go as far as 45° N latitude before heading E and SE.

On the third day we had reached 27° N and could hold 015° true course. Thus, a slight wind change was occurring. It wasn't possible to obtain a celestial fix, however, because of morning and evening cloud cover. The passage, so far, had been beautiful, and the steady winds had held us on course with very little needed attention to the helm.

With four persons aboard, the watches were divided into two-hour periods, so we had much free time which was utilized by reading and playing backgammon. So that each of us wouldn't have the same daily watch schedule, I extended two of the watch periods to three hours: 1000 to 1300 and 1300 to 1600. During our voyage, we had learned that everyone was active during these times, and watches were usually relaxing. This procedure seemed to work quite well.

On the fourth day out we could hold a course of 030° true. The weather was fair, and the seas were kind. Our supply of ice would last four or five more days, so we were still having fresh rations to prepare our meals. We enjoyed bacon and eggs for breakfast, and this night we were to have steak.

By the late afternoon our course was 040° true. We watched four albatross that had been following us all day. As evening approached, the pastel colors of the sky, sea, and clouds developed in a regular progression of spectacular changing colors. Soon we would see Venus in the western sky and then look for Arcturus and Spica in the SE. Then, other familiar beacons would appear. As we prepared for the evening celestial sights, I contrasted this passage with the one to Hawaii and wished that Nan could be with us now. Our perfect fix placed us forty-seven miles NE of our D.R. position. Thus, a feeling of peace and pleasure settled over us, as we had popcorn at 2200 before the off-watch crew retired.

By 8 June we had reached a latitude of 31° 08.0′N, but still hadn't picked up the westerlies. The engine was run for two hours to charge the batteries, after which the nut on the packing gland was again tightened. For a day the winds changed, and we were able to make a course of 060° true, but before long, we were back to about 030° true.

A new interest was added to our daylight watches. The alert helmsman would spot a glass Japanese fishing float. Then, excitement would prevail as we maneuvered for a position to pick it up. If a float had a line attached to it, collection would be easy, but many times a glass ball would slip from the fingers just at the critical moment when it was lifted. Once obtained, our prize would be scrubbed with a stiff brush and soap to clean the growth of algae and barnacles. Many also had small crabs clinging onto the surface.

The temperature of the night was such a contrast to the South Pacific that we were now wearing thermal underwear, coats, foul weather gear, and a watch cap. The days, however, were sunny and warm. We had reached 33° 43.0′N latitude, but were able to hold a course of 070° true, which was most encouraging. Our longitude was 148° 30.0′W, indicating a good eastward progress.

Since we were back on canned rations, meals became interesting as various "cooks" improvised to create new gastronomic delights. For breakfast, for example, Russ made pancakes using beer instead of water with the mix. Surprisingly enough, an interesting yeasty flavor was pleasant to the palate. A topping of butter, powdered sugar, and a little lemon made them even more interesting.

Our 12 June celestial fix showed that we were only two miles off our D.R. This was truly accidental! A freighter heading NW was seen just off our port bow. She slowed in the water until after we passed, before steaming into the distance. Again, I felt security in knowing that captains of these large ships observe small boats. Later in the night another freighter crossed our stern heading SE. Our running lights were on, so I assume we were observed, although this time, no recognition was given. Then, in the morning another freighter heading NW remained dead in the water until after we passed. This Esso tanker was obviously empty, as she floated high in the water.

The wind abated and we were suddenly on a quiet, undulating surface. The sails flopped and became useless. We had entered a high pressure area, which fortunately wasn't very extensive. After a day and a night of powering, we found the wind. This called for a celebration. Brad made a cake. Karl, who was not known for his culinary efforts, was going to fix dinner. He began to gather cans and planned to mix chili beans, tomatoes, corn, peas, and carrots together. At the realization of what was about to occur, Brad, Russ, and I changed the menu. Brad made garlic bread, Russ heated chili beans only, and Karl served. When Karl admitted later that he detested cooking, we could all believe him! Perhaps Karl would be better suited to the dishwashing detail, I thought

For the next three days the wind picked up, but so did the seas. Whitecaps and waves slowed us down. After our luxurious, comfortable beginning, I wondered if this would be a repeat of our voyage to Hawaii. One night, about 1700, I noticed that the head wouldn't flush. *There are no plumbers at sea.* How true! Brad was on watch, Russ was fixing supper, so that left Karl and me to attend to the problem. Eventually the entire bowl had to be removed in order to check all the lines. About 14 inches in the pipe, I found the problem. Kleenex and toilet tissue had accumulated, forming an effective plug. Everything was washed, reassembled, and installed, but a leak developed in the junction of the outlet pipe and the pump. Since it was 1900, I decided to locate the problem in the morning. Karl got sick, but a couple of trips to the fresh air soon effected a cure.

I relieved Brad on watch and Russ served supper of corn

bread, stew, Brussels sprouts, peas, and coffee. Although the required activity hadn't been very pleasant, the exercise did stimulate our appetites.

The next morning, Brad and Russ put a new gasket on the head outlet, which stopped the leak and put this vital function in operation once again.

On 16 June we were still in heavy weather and making slow progress. I had just come off watch and was warm, comfortable, and sound asleep when Russ woke me at 0800. Several jib snaps had worn through and several slides on the main needed resewing. When this happens, and one of them breaks, the additional strain on neighboring ones causes them to let go also.

We all went above, and I started the engine as the jib and main were furled. I set the autopilot to keep us into the wind and adjusted the engine to 1200 rpm, to keep us barely moving through the whitecaps and heavy swells. Then, Brad and I worked on the foredeck replacing the worn jib snaps while Karl and Russ sewed the slides on the main. While working on the jib, I noticed that there was no cotter pin in the clovis holding the forestay in place. How lucky we've been! If there had been a cotter pin here, how could it have come out? Why didn't one of us notice this when we checked all the fittings before leaving Honolulu? Did Ala Wai Marine, Ltd. put one in? If so, why wasn't it properly bent to hold it in position? These questions will never be answered, and again came the realization that this was also negligence of the captain! A stainless steel cotter pin was put in place and bent—even before we completed replacing the jib snaps.

By 1000 we were sailing again, using only the jib and mizzen. It was really getting cold, and unusual breaking waves occasionally splashed the helmsman, which didn't help. The sun remained hidden behind dark clouds.

The sun, fair skies, and calming seas were a welcome change after three days of rough weather. However, it remained cold, even below, so I built a fire in the Cole-stove and kept briquettes burning in the mornings and evenings. It was very pleasant to feel the warmth of the cabin when a stint at the helm had been completed. As the wind became more favorable, we could easily hold a course of 070° true.

On the evening of 18 June we picked up two San Francisco radio stations: one a rock-and-roll station, the other a religious one, reciting Hail Marys. About the same time, a Sea-Train Line freighter traveling NW crossed about five miles in front of us. These two unrelated events raised our spirits, and Karl decided to fix dinner. Not wishing to thwart his initiative, we agreed. Our one-bowl dinner that evening was made by combining two cans each of sauerkraut, potatoes, and Vienna sausages in a pan and heating the mixture on the stove. Brad had made a chocolate cream pie with a graham cracker crust for desert.

In the early morning hours a Russian freighter, *Republica,* crossed our bow heading west. She came very close to our port side, slowed down, and then resumed speed after we had passed. Our D.R. position was latitude 38° 40.0'N and 132° 28.0'W. We were 510 miles from Santa Cruz. Two more freighters passed us that day, so we would have to be alert as we approached the California coastal waters.

On the morning of 22 June, Karl was to experience his first emergency at sea. I relieved Brad at 0600 after which he was below charting our night's progress. There was a sharp noise as the forestay parted, and then I yelled, "Everyone on deck." We got the jib and main down and temporarily secured the jib boom on the foredeck before we had time to inspect the problem. The forestay turnbuckle had cracked through the threads and parted.

Having had a similar emergency months before, our action was efficient and effective. Russ took the wheel as we ran with the seas. Brad got a line around the forestay, through the anchor roller, and around the anchor winch. The winch, then, was used to make the forestay taut before temporarily securing the line. A block and tackle attached to the forestay and to the bow fitting with shackles then replaced the temporary line. For a safety precaution, a 3/8-inch stainless steel cable was attached to the forestay and to the bow fitting with a niko-press. It seemed right at first, but after discussion, it appeared to us that if the block and tackle should let go, the stainless steel cable would surely snap under the sudden strain. In fact, the more we talked, the less secure we felt about this corrective measure.

Brad and I went below to get some anchor chain, from which we cut a three-foot length. When this was attached, with shackles, to the forestay and to the stainless steel bow plate, the block and tackle was backed off a bit to let the chain take the strain. Now the emergency had passed. Karl lashed the jib boom to the deck, as Brad and I bent the jib onto the forestay. Many boats do not use a boom for the jib, and such an arrangement is called a "loose-footed jib." Our jib sail, although it wasn't cut for such a condition, filled and pulled nicely. I decided then that the next time I go cruising, there must be two headstays on the boat.

The entire repair had taken four hours and, in that duration, the wind had strengthened considerably, so I felt content to travel at five knots under the jib and mizzen alone, keeping the mainsail furled. At 1030, poor Karl was just coming down to go to bed when I had to remind him that it was his watch time.

In the evening, we were able to get a good RDF (radio direction finder) fix using Eureka, the Farallons, San Francisco, and Santa Cruz. We were able to get Monterey (KMBY), but not strong enough to use: latitude 37° 45.0′N and longitude 125° 50.0′W. Thus, we were 188 miles from Santa Cruz.

At 0330 on the morning of 23 June, a flashing light brought us all on deck in spite of the predawn cold. Identifying the beacon as Montara Point was verified by a fix on a Santa Cruz radio station. Only sixty miles to go. My "prediction time calendar" read "0 days," so we were going to make our landfall in the time estimated a month ago in Honolulu.

Much later, through a light coastal fog, we saw the coastline of California. The thrill of identifying Pigeon Point cannot possibly be put into writing, but even now, as I think about that moment, a strange tingling of anticipation overcomes me. I suspect there will be three others who will experience the same sensations when that moment is recalled.

My "dumb" radio blew a fuse when I tried to effect a radio-telephone call to let Nan know that we were actually going to arrive on time.

At 1435 we approached Davenport and made one more long tack to clear the point. At 1715 Pacific Daylight Savings Time, *Pagurus II* cleared the channel entrance and slid into her home berth, D-13, at Santa Cruz, California. It seemed as though the docks wouldn't hold still, as I made my way up to the telephone on shore. A short prayer of thanks, and then: "Nan, we're home."

APPENDIX

PAGURUS II NEWSLETTER #1
21 February 1974

As the time for planning is really here, there are many loose items to be organized to avoid omission. For example, my addresses and telephone numbers of the members of the expedition are not complete. It would be well, also, if you would each check over my rough attempt at a biography to correct or add data pertinent to our purpose.

I am currently preparing lists of: 1) items to have done; 2) spare parts; 3) suggested immunizations, health, and dental records; 4) supplies; 5) food; and 6) a suggested clothing list. We'll have to get together for discussions on a practical reference library. We have a growing collection of paperbacks!

At any rate, when you look over this beginning ship's list, will you note any suggestions for additions of items or categories and let me know what they are. Russ, for example might like to determine the distances between the listed Ports of Call. Dave might have a suggestion for one we missed. Irve might like Bora Bora (which we will probably see anyway).

See you soon,
Al Towle

EXPEDITION OF THE *PAGURUS*

Dates: June or July 1974 to March or April 1975 (anticipated)
Planned Route: (subject to change depending upon time)

San Diego
Acapulco
Isla del Coco
Galápagos Archipelago
Marquesas Islands—fuel in Nuku Hiva
Tahiti and Society Islands—fuel & supplies
Honolulu
San Diego
Santa Cruz

Crew: There will be no pay or fees charged, as this venture is of joint interest. All members of the expedition are scientifically trained.

Al Towle [address and telephone number]

Sailing since 1948 when in the navy at Pensacola, Florida

Have owned two boats

Owned *Pagurus II* since 1966

Cruise to South Channel Islands and Baja California every Summer

Taking celestial navigation course from Capt. Todd at California Academy of Science (3rd time)

Irwin Thornton [address and telephone number]

Retiring high school science teacher

Owns a Winslow 30-foot ketch and has had it about 13 years

Sailing experience for about 40 years

I have known him for 20 years

Dave Feldman [address and telephone number] Student working on M.A. degree at San Francisco State University

Has had some sailing experience

Russ Mabardy [address and telephone number]

Taking celestial navigation course from Capt. Todd at California Academy of Science

Has had a course in sailing at San Francisco State University

Has sailed from San Francisco to Seattle as a crew on another boat

Student working on M.A. degree at San Francisco State University

Has assisted on *Pagurus II* when other students from classes at the University have gone collecting on Monterey Bay

Estimated cost of this trip to me:
$10,000.00

Plans and Project

1. Travel to the Galápagos archipelago and selected South Seas Islands for survey of sipunculan fauna and to assess the reproductive state, study the coelomic cells, and to prepare tissue for biochemical analysis.

The value of research on the Sipuncula is that it will add to our knowledge on distribution of the phylum and provide more information on the morphologically diverse cells in the coeloms of these animals. Related problems, naturally, offer topics of investigation for my students back at the University and thus, problems of research for graduate students.

This area of research is not something new to me. I have been working with these animals for 16 years. In fact, in 1970, I was invited and attended an International Symposium in Yugoslavia to present a paper on the Sipunculida in Monterey Bay. I have also looked for these animals (without success) in Antartica. Some larger forms do exist in Arctica, although I have neither collected nor studied them there.

2. Two of my students (M.A. candidates) will be working on faunal distribution as part of their research for the advanced degrees.

Preplans and justification

In anticipation of returning to the Galápagos in my sailing ketch, I visited the islands during the Summer, 1972, with a U.C. Extension Cruise. Therefore, I have identified locations where I would like to return for study, and have identified secure anchorages.

Sipunculans live in tropical, temperate, and polar seas. Some species have a wide geographical distribution. Some, like *Siphonosoma cumanense, Phascolosoma nigrescens,* and *Cloesiphon aspergillus* are associated with coral reefs and are circumtropical. One sipunculan *(Golfingia margaritacea)* is claimed to be bipolar (Fisher, 1920; Theel, 1911; Stephen, 1941). It is not known whether there is a continuous distribution of the species between the polar regions. Some sipunculans are more widely distributed than formerly believed (Murina, 1957, 1958, 1961, 1964, 1968).

The taxonomy of this phylum has been recently reviewed by Stephen and Edmonds (1972), although the reproductive habits and physiology have not been extensively studied. Coelomic cell types have been studied by Towle (1973), Green (1973), Ohuye (1961), Ochi (1973), Florkin (1933, 1934), and Craw (M.A. thesis now completed under Towle).

Sipunculans have not been studied from the Galápagos Archipelago, nor can references be found where they have ever been discussed.

It is my desire, therefore, to: 1) examine the islands for sipunculans and to identify them; 2) assess the reproductive state of the populations; 3) examine the coelomocytes; 4) preserve specimens for biochemical assay on returning to San Francisco (if arrangements can be made to remove specimens).

SELECTED REFERENCES

1. Florkin, Marcel. 1933. Recherches sur les Hémérythrines. *Arch. Internat. de Physiol.* 36: 247–328.
2. Fisher, W. 1920. Gephyreen der antarktischen and sub-antarktischen Meere. *Dts. Sud. Pol. Exped.* 16, Zoologie (8): 409–430, figs 1–4.
3. Green, S.A. 1973. The annual reproductive cycle of *Phas-*

colosoma lurco. Proceedings of the International Symposium on the Biology of the Sipuncula at Kotor, Yugoslavia, June 18 to June 25, 1970.

4. Ochi, O. 1973. An electron microscope study on the coelomic cells of some Japanese Sipunculoidea. Proceedings of the International Symposium on the Biology of the Sipuncula at Kotor, Yugoslavia, June 18 to June 25, 1970.
5. Ohuye, T., O. Ochi, and J. Miyata. 1961. On the Morphogenesis and Histochemistry of the "urn" found in the Coelomic Fluid of a Sipunculid, *Phascolosoma scolops. Mem. Ehime Univ.* Section II (Science) Ser. B (Biology) IV (2).
6. Stephen, A.C. 1941. The Echiuridae, Sipunculidae and Priapulidae collected by the ships of the Discovery Committee during the years 1926–1937. *Discovery Rep.* 21: 235–260, pls 7,8.
7. Stephen, and S.J. Edmonds. 1972. *The Phyla Sipuncula and Echiura.* The British Museum (Natural History). London.
8. Theel, H. 1911. Priapulids and sipunculids dredged by the Swedish antarctic Expedition 1901–1903, and the phenomenon of bi-polarity. *K. svenska Vetensk. Akad. Handl.* 47: 1–36, pls 1–5, 8 figs.
9. Towle, A. and A.C. Giese. 1966. Biochemical Changes During Reproduction and Starvation in the Sipunculid Worm *Phascolosoma Agassizii. Comp. Biochem. Physiol.* Vol. 19, pp. 667–680. Pergamon Press Ltd. Great Britain.
10. Towle, A., and A.C. Giese. 1967. The Annual Reproductive Cycle of the Sipunculid *Phascolosoma Agassizii. Physiol. Zool.* 40: (3) U. of Chicago.
11. Towle, A. 1973. Cell types in the coelomic fluid of *Phascolosoma Agassizii.* Proceedings of the International Symposium on the Biology of the Sipuncula at Kotor, Yugoslavia, June 18 to June 25, 1970.
12. Murina, V.V. 1957. Sipunculids collected on the first trip of the Complex Antarctic Expedition of the 'OB in 1956. *Zool. Zh.* 36: 992–996. 1 map (Russian with English summary.)
13. Murina, V.V. 1958. On the systematics of two closely related species of deep-water Sipunculids of the genus

Golfingia, according to the materials of the 'Vitjaz' Expeditions in 1949–1955. *Zool Zh.* 37: 1624–1634, 1 fig. (Russian with English Summary).
14. Murina, V.V. 1961. The geographical distribution of abyssal Sipunculida. *(Phascolion lutense* Selenka.) *Oceanol.* (Acad. Sci., S.S.S.R.). 1961. (I): 140–142, 1 fig.
15. Murina, V.V. 1964. On the sipunculid fauna of the Mediterranean. *Trudy sevastopol. Biol. Sta.* 17: 51–76, 20 figs.
16. Murina, V.V. 1968. Occurrence of abyssal sipunculids in the Atlantic Ocean. *Zool. Zh.* 47: 195–198, 1 fig.

Description of Pagurus II
(Winslow-designed Blue-water Diesel Offshore Ketch)

LOA	41′9″
LWL	30′10″
Beam	11′0″
Draft	6′0″
Displacement	26,000 lbs.
Keel Ballast	10,000 lbs, iron

Sails all dacron

	sq. ft.	age
mizzen	175	3 yrs.
main	394	2 yrs.
jib	288	2 yrs.
jenny	600	new
stormsail	90	heavy canvas

Planking, mahogany
Frames ipol
Fastenings, copper
Keel, stem, deadwood, floors: yacal
Mahogany trim
Rudder teak
Cabin sole teak

Decks and cabin top: fiberglass covered marine plywood
Spars hollow spruce
Rigging stainless steel
Sheet and halyard winches
Synthetic running rigging
Stanchions with plastic covered lifelines
Plastic covered foam mattresses (6)
Fixed bilge pump (Whaler 15 gpm) + automatic
Automatic alarm system
Bronze seacocks
Tin-lined copper fuel and water tanks

Engine
    Perkins 4–107 diesel
    fuel 80 gallons + 50 deck load
    water 100 gallons + 50 deck load

Bilge pumps
1. Fixed Whale gusher (15 gpm)
2. Jabsco water-puppy, automatic

Dinghies
1. 10′ Avon rubber dinghy and oars
2. Boston Whaler and outboard (to be purchased this spring)

Engine accessories
    HEP to prevent electrolysis
    Aqualarm (for bilge water, oil pressure, and engine temperature)
    Silicon diode for charging batteries

Batteries
    2 Trojan, 180 amp-hour batteries in 2-volt cells, 1 year old

Navigational gear
    Apelco 101 W UHF ship to shore radio (WZW 3905)
    Apelco recording fathometer
    Bendix Navigator 555 Radio Direction Finder

Bendix Model #14 Auto Pilot
Wind vane (to be installed this spring)
Danforth-White Constellation compass
Danforth-Corsair compass
Waltham boxed and gimballed chronometer
All approved navigational lights
Two sextants
All appropriate sailing charts
Pilot charts for North Pacific and South Pacific
Star identification and finder
Navigational Books: Mixter, G.W. *Primer of Navigation,* 5th ed.
Hill, John II, R.F. Utegaard and Gerald Riordan. 1969. *Dutton's Navigation and Piloting.* 12th ed.
2 nautical almanacs
2 Ageton, H.O. #211 tables
4 parallel rules
4 dividers
4 triangles
Position plotting sheets for appropriate latitudes

Scientific equipment
diving gear
hand lenses
microscope, compound
microscope, dissecting
collecting vials
formaldehyde
reference books
trawl net
photographic equipment for above and below water

Safety Gear
2 Flare pistols and flares (night, day, and parachute)
2 horseshoe life preservers
5 life jackets, Coast Guard approved (2 years old)
2 safety harnesses
1 bosun's chair
2 strobe safety lights (man-overboard type)
1 12 volt spotlight

2 whistles
Stainless steel bow and stern pulpit
Stainless steel safety lines.

First aid kit
To be complete and made up by Karl Wolf, M.D., of Manteca and Darrell Sedgwick, M.D., of Capitola

PAGURUS II NEWSLETTER #2
20 March 1974

To the crew:

I saw First Mate Irwin Thornton this week-end, and he has his retirement papers in. He has also related the cruise plans to many envious colleagues. Also, I know Dave will be interested to know that Irve has put down another batch of his home-grown olives in rock salt. The first batch passed the taste test!

Russ and I have been plugging along with celestial navigation, and he informs me that he thoroughly understands the various complicated time manipulations, so when we cooperate on the plotting, we should obtain good fixes. My Waltham chronometer has been set in motion and will go in for a check next month.

I have written to Craig MacFarland, Director, Charles Darwin Foundation for the Galápagos Isles to ask if we could bring anything down that he needed or if he can help us obtain special clearance for our work. I hope my next newsletter will convey a response.

Dave Feldman's address was omitted from Newsletter #1. It is:———.

We'll have to get together for an evening sometime (perhaps in May), and I shall show a few slides of the Galápagos, its flora, fauna, and lava mountains.

Included in this Newsletter are lists of items to which you should direct your attention. You will probably want to use this for a checklist.

See you soon,
Al Towle

Essential items:

Passport valid through 1975

World Health Certificate (Expiration date beyond 1975) Smallpox vaccination, Tetanus, Typhoid immunization

Traveller's checks and personal checks

Credit cards

Spending money (include small denominations of U.S. currency)

Reserve money to finance your own way home should an emergency arise

Personal effects:

Wash cloths(2)

Towel (2)

Bar soap (12)

Tooth paste (6 large)

Shaving gear

Sunglasses

Insect repellent

Toothbrush

Suntan lotion (with good U.V. filter)

Sunburn ointment

Chapstick (about 5 or 6)

Personal equipment:

Pen, pencil, notebooks, paper

Camera and film (You may want to store your film in dehumidifying plastic bags available at most photography stores)

Plastic bags to protect camera equipment when going ashore

Exposure meter—if your camera doesn't have one built in

Spare batteries for camera or flash attachments or light meters.

Lightweight canteen (We should have at least 2 aboard)

Flashlight (non-ferric) with fresh replacement batteries

Pocket knife (Swiss Army type is good)

Sleeping bag (one that is not too heavy, and down that will dry quickly if it gets wet)

Clothing:

Hiking shoes (1 pr.)
Deck shoes (2 prs.)
Bathing suit (1)
Walking shorts (2 pr.)
Washable trousers (4 pr.)
Underpants (6)
Tee-shirts (6)
Drip-dry short sleeved shirts (2)
Long-sleeved cotton shirt—such as kahki (1)
Socks (6 pr.)
Light windbreaker (1)
Washable shore outfit (1)
Heavy coat (one that dries quickly) (1)
Thermal underwear (1 set)
Hat or cap (1)
Sweater or sweatshirt (1)
Light (foldable) raincoat (1)
Foul weather pants
Foul weather jacket]You might check Rocket Surplus on Broadway in Redwood City.

Food:

As on any excursion, food is an important morale booster. The following is merely a basic list of items I have considered for provisions. Will you please look it over and let me know what you would like added.

[Here the supplies were listed under the following categories:

Perishables, Dried Foods, Canned Goods (meats), Canned Goods (vegetables), Canned Goods (fruits), Canned Goods (miscellaneous), Meats (fresh), Jars and Bottles, Miscellaneous, Spices, Wines, Beer, Liquor, Paper Goods, Soaps.]

Galley Equipment:

Inventory to be taken later

Medical and First Aid Supplies:

Inventory to be taken later

Medical and Dental Check-Up:

Be certain to have a physical and a dental check within the next two months. Any pertinent medical information or allergies, please let me know immediately.

[Following this information in Newsletter #2, there was a map of the Galápagos Archipelago, and information about the Fondation Charles Darwin pour les Galápagos. Then, there was a request for information regarding the name, address, and phone of nearest kin, passport number, expiration date, and dates of immunizations.]

PAGURUS II NEWSLETTER #3

23 May 1974

As the time draws near the end of the 1973–74 school year, we can begin thinking about the items to be completed in June. Plans call for varnishing the sides, painting the decks (with sand) to roughen up the walk-ways, tuning up the shrouds, tuning up the engine, checking all spare parts, and then removing everything from below to begin proper stowage.

Stowing will occur during the first two weeks of July. Dave has agreed to act as storekeeper, so he will inventory all the items as they are stowed in various lockers. Therefore, if you can't find something, ask Dave.

Russ is putting together the final list of navigational tools, books, and charts, so they will be ordered next week. You may all be interested to know that I passed the two-part final on the celestial navigation course. Russ will pass his final next week, so we will have two qualified navigators aboard. The rest of the crew will study celestial navigation at regular intervals on the cruise.

First mate Irve Thornton is about to become a retired teacher. He has been busy painting and rigging his boat so it will withstand the rigors of being left alone in Santa Cruz during the absence of her captain.

A set of signal flags and flags for the countries of Mexico, Costa Rica, Ecuador, and France have been ordered. Also the needed spare-parts list is shrinking!

The Riebandt wind-vane steering has been ordered and this weekend, Nan and I went to Dana Point to make arrangements for stopping there for a few days while the wind vane is being installed.

For the ship's records, will each of you please send me the following information right away:

Name Age
Passport number
Place and Date of Birth
Address
Physical description (As on the driver's license)
Name of nearest kin
Address of nearest kin

This is necessary, as I must type them, and have several copies aboard to be used for Port Entry Clearances.

We will have a get-together dinner during the month of June, and I can show you a few slides of the flora and fauna found on the Galápagos.

See you soon,
Al Towle

LIST OF SHIP'S PAPERS CARRIED ABOARD

1. Correspondence
2. Pagurus Newsletters
3. Information for Scientists
4. Proposals for Study
5. Sabbatical Leave Authorization
6. Equipment Use Authorization
7. *Pagurus II* Specifications
8. Certificate of Ownership
9. Spanish Entrance Papers
10. French Visa Information for Crew
11. Crew List and Requests to Sail
12. Galápagos Information
13. Manifest
14. Medical Supplies

15. Food Supplies
15. Food Supplies
16. Riebandt Vane Steering Information
17. Voyage of *Pagurus II* requests
18. Navigation Charts Aboard
19. Spare Parts List
20. Instruction for CROTON Terrestrial Watch
21. Passports and Visas
22. WWV Broadcast Information
23. Traveler's Checks numbers for Crew

CONTENTS OF MEDICINE CHEST OF YACHT *PAGURUS II**

1. Adrenalin. Injectable. Concentration 1:1,000. To be used in situations of severe allergic reactions or shock. Dose: .2 to .5 cc.
2. Antivert. 25 mg. To be used for vertigo, seaskickness. One tablet 4–6 times per day as necessary.
3. Ammonia. Breakable capsules. Use for fainting spells or dizziness.
4. ASA (aspirin). For aches and pains. One to two tablets four times per day.
5. ASA w/Codeine 30 mg. for severe pain. One to two tablets up to four times per day.
6. Bactrim. To be used strictly for urinary tract infections. Dose: 2 tablets twice a day for ten days.
7. Benadryl. 50 mg. capsules. To be used for combating allergic reactions, rashes, or at times nausea and vomiting, or may be used for a sleeping tablet. Dose: May be used from 1 to 4 capsules per day as necessary.
8. Betapen. This is a general penicillin. Tablet, 250 mg. Use for general infections. Dose: 1 tablet four times per day.
9. Camalox. This is an antacid. May be used for upset stomachs. Use 1 tablet four times per day and at bedtime

*Note: Since I am not an M.D., I take no responsibility for the accuracy of the information contained in my Medical Chest. This material is informational only. Any serious cruising person may consult a physician regarding various items.

or 1 or 2 tablespoons up to four times per day, as necessary for upset stomach.

10. Conar-A. This is a cough tablet. 1 tablet four times per day. May also be used as a general cold tablet.
11. Combid. These capsules may be used for severe nausea or vomiting or cramping, abdominal pains, and diarrhea. One tablet twice a day.
12. Co-Pyronil. This is a decongestant. May be used for allergic hay fever, rhinitis, or cold congestion. One tablet three times a day.
13. Coricidin "D". General cold tablet. One tablet up to four times a day.
14. Darvon Compound 65. For moderate pain. One capsule up to 4–6 times a day.
15. Decadron Injectable. Concentration 4 mg./ml. For severe allergic reactions or in states of shock. Dose: One ml. injectable, as much as twice per day.
16. Dimetabs. Decongestant. One tablet twice a day.
17. Dramamine. To be used for seasickness, dizziness. One tablet up to four times per day.
18. E-Mycin. This is a general antibiotic to be used for those members of the crew who are allergic to other antibiotics, i.e., penicillin. General dose: One tablet four times per day
19. Erythrocin. Again, another erythromycin product. Strength 250 mg. For general infections. One tablet four times per day.
20. Elixir of Terpin Hydrate s/Codeine: One teaspoon every 4 hours as necessary for cough.
21. Fiorinal. One tablet to be used every 4 to 6 hours as necessary for tension headaches.
22. Gantrisin: 500 mg. Two tablets 4 times a day for urinary tract infections. In the event of other antibiotics being used up, this may be used as a general antibiotic.
23. Hyper-Tet. Immune globulin. This can be injected at 250 cc. per dose in the event of a serious or extensive injury involving exposure of large areas of musculoskeletal tissue or severe penetrating intraabdominal injuries. Only one dose need be given.
24. Ilosone. 500 mg. This again is an Erythromycin product. One tablet four times a day for severe infections, for

those members of the crew who are allergic to penicillin.

25. Librium. 10 mg. It is a general tranquilizer. One tablet four times a day maximum.
26. Lomotil. To be used for diarrhea. One tablet up to four times a day as necessary for diarrhea.
27. Loridine. This is an injectable antibiotic for severe infections in a patient who cannot take oral antibiotics. 1 cc. equals 500 mg. Inject 1 cc. three times a day.
28. Lidosporin Otic: To be used for external infections of the ear. Use two drops four times per day as necessary.
29. Maalox. This is an antacid for general stomach upsets. Use as Camalox above.
30. Marzx. Thus is an antibronchospastic agent which prevents allergic wheezing and allergic shortness of breath. One tablet up to four times per day.
31. Nebs. These are analgesics similar to aspirin without the gastric side effects of upset stomach. Use as aspirin.
32. Neosporin Ointment. This is a topical antibiotic agent for surface wounds.
33. Norflex. This is for back sprains and muscle spasms. Use one tablet four times per day.
34. Nupercainal Ointment. This is a topical anesthetic agent for surface wounds.
35. Nupercainal Suppositories. For rectal pain and hemorrhoids.
36. Phenergan-D. To be used in association with Antivert or Dramamine for severe seasickness. One capsule four times per day.
37. Prednisone. 5 mg. To be used for significant allergic reactions. One tablet 3–4 times per day for no longer than four days. Use this medication for any severe allergic reactions, not only involving the skin, but also the respiratory tract, i.e., asthmatic-like conditions. Also suitable in hives and severe cases of laryngitis with laryngeal swelling.
38. Robicillin VK. This is a general penicillin. 22 mg. per tablet. Used for general infections. One tablet four times per day.
39. Sulamyd. This is an ophthalmic solution for eye infections or any eye injury. Two drops four times per day.

40. Surfacaine. This is a topical anesthetic agent for general surface wounds.
41. Tetrex. 250 mg. General antiobiotic. One capsule four times per day.
42. Tetrex. 500 mg. For severe infections. Again, generalized infections. One capsule four times per day.
43. Tetanus Toxoid. Booster shot of this to be given with any severe injury and should be given in association with Hyper-Tet in those instances as outlined under Hyper-Tet. Dose: .5 cc. intramuscular.
44. Tinactin. This is a topical agent for fungal infections, i.e., athletes foot and ringworm of the body or head.
45. Valium. 5 mg. This is a general tranquilizer. One tablet three times per day.
46. Valisone. This is a topical cream for allergic type rashes or may be used as a general ointment for rashes in which the skin has not been broken.
47. Vosol Otic. This is a solution to be used for canal infections of the ear. Two drops four times per day.
48. Xylocaine (Lidocaine)

Additional medications:

Crysticillin (Penicillin G)
Prostaphlin (Sodium Oxacillin capsules 500 mg.
Quintess—for diarrhea
V-Cillin K (Potassium Phenoxy-methyl Penicillin—250 mg.—400,000 units)
Fluonid cream for topical application
Fever Thermometer
Alcohol sponges
Syringes
Needles
Minicatheters
Cavit (temporary dental fillings)
I.V. Injection apparatus
Finger Splints
Bottles of I.V. saline solution
Ace Bandages (assorted sizes)
One copy of the *Merck Manual*

PAGURUS II FOOD STORES
(aboard when leaving Santa Cruz)

Perishables

20 lbs. Potatoes
20 lbs. Carrots
12 heads Cabbage
3 bunches Celery
12 doz. Eggs
6 loaves Bread
12 pkgs. English muffins
Oranges, grapefruit, lemons, limes

Dried Foods

Oatmeal, 3 boxes
Powdered milk, 12 boxes
Powdered chocolate, 6 boxes
Pancake mix, 12 boxes
Cake mixes, 12 boxes
Flour, 20 lbs.
Sugar, 20 lbs.
Salt, 12 boxes
Crackers, 12 boxes
Rye-crisp, 6 boxes
Biscuit mix, 6 boxes
White rice, 5 lbs.
Brown rice, 5 lbs.
Spaghetti, 10 lbs.
Potato flakes, 12 boxes
Split peas, 5 lbs.
Mushrooms, 12 pkgs
Cookies, 24 pkgs.
Raisins, 10 lbs.
Prunes, 10 lbs.
Figs, 5 lbs.
Soups, 48 pkgs.
Onion, 12 boxes
Coffee (instant) 24 small jars
Coffee (regular), 24 2-lb Max-Pax
Tea (instant), 4 dz jars
Pream, 6 med. bottles
Cup-A-Soup, 48 pkgs.

Canned goods—Meats

Bacon, 48 1-lb cans
Ham, 6 cans
Corned beef, 2 cases
Luncheon meats, 1 case
Swedish meatballs, 1 case
Chicken (whole), 24 cans

Ravioli, 2 cases
Spaghetti and Meatballs, 1 case
Chili and beans, 2 cases
Roast beef hash, 1 case
Corned beef hash, 1 case
Cheese rarebit, 1 case
Tuna, 2 cases
Vienna sausages, 1 case

Canned Goods—Vegetables

Asparagus, 1 case
Beans, green, 1 case
Beans, kidney, 1 case
Beets, sliced and julienne, 1 case
Carrots, 1 case
Corn, 1 case
Garbanzos, 12 cans
Sauerkraut, 12 cans
Tomato juice, 2 cases, large cans
Tomato sauce, 1 case, large cans
Soups, mixed, 2 cases
Peas, 1 case
New Potatoes, 2 cases

Canned Goods—Fruits

Orange and grapefruit juice, 1 case, large cans
Pears, 2 cases
Applesauce, 2 cases
Pineapple, 1 case
Fruit salad, 1 case
Lemon juice, 4 cans
Lime juice, 4 cans
Orange juice, 4 cans

Canned Goods—Miscellaneous

Milk (evaporated), 1 case
White sauce, 12 cans
Olives (mixed), 1 case
Nuts, 2 cases
Smoked oysters, 1 case
Canned butter, 1 case

Meats (fresh)

Salami, 12
New York Steak, 8
Sirloin tip roast, 3
Smoked pork chops, 16

Jars and Bottles

Catsup, 12 bottles
Cheese spreads, 6
Honey, 2 jars
Jam, 12 large jars
Mustard, 6
Peanut butter, 12
Pickles, 4
Corn Oil, 8 qts.
Syrup, 12 qts.
Vinegar, red, 2 qts.
Vinegar, white, 2 qts.

Miscellaneous

Cheese, 10 lbs
Imo sour cream, 3 pts.
Margarine, 6 lbs.
Baking powder, 2 cans
Cornstarch, 1 pkg.
Cornmeal, 6 pkgs
Crisco shortening (small) 1 can
Chocolate bars, 1 case
Imitation sour cream, 6 cans

Spices

Pepper, 2 large jars
Paprika, 1 can
Mustard, 1 can
Curry powder, 2 jars
Tarragon, 1 jar, flakes
Marjoram, 1 jar, flakes
Oregano, 1 jar, flakes
Cinnamon, 1 large can
Garlic salt, 2 jars
Horseradish, 1 bottle
Salad dressing mixes, 6 pkgs
Soy sauce, 2 bottles
Tabasco, 2 bottles
Worcestershire sauce, 1 bottle

Spirits

Wines, 4 cases
Beer, 2 cases
Hard Liquor, Assorted

Paper Goods

Toilet tissue, 48 rolls
Paper towels, 24 rolls

Paper napkins, 3 family packs
Paper plates, 6 dz. pkgs
Paper cups, 3 pkgs
Tissues, 4 boxes
Aluminum foil, 3 rolls
Saran wrap, 3 rolls

Soaps

Joy, dish soap, 12 bottles
Lysol spray, 6 cans
Scat, 1 can
Scouring powder, 12 cans, plastic

Additional supplies were purchased along the way.

NAVIGATIONAL CHARTS

British Admiralty

992	Archipel des Tuamoto
998	Ile Makemo to Ile Tahiti
1060	Huahine to Maukita
1103	Tahaa and Raiatea
1107	Iles de La Societe
1175	Anchorages in the Archipel Des Tuamotos
3045	Enderburg Island to Christmas Island
3664	Plans in Archipel des Tuamoto

Coast and Geodetic Survey

4100	Island of Kawaii
4103	Hilo Bay
4109	Island of Oahu Honolulu Harbor
4110	Island of Oahu
4115	Island of Hawaii
4116	Hawaii to Oahu
4117	Oahu to Niihau
4120	Channels between Oahu, Molokai & Lanai
4130	Channels between Molokai, Maui, Lanai, & Kahoolewe

4131	Southeast Coast of Oahu
4132	South Coast of Oahu
4133	Oahu, Ahua Point to Barber's Point
4140	West Coast of Hawaii Cook Point to Upola Point
4164	Kailua Bay
4167	Kawaihae Bay
9000	San Diego–Aleutians–Hawaii

Naval Oceanographic Office

526	Pacific Ocean Central Part
527	Pacific Coast of Mexico and Central America
528	Pacific Coast of U.S. and Mexico
620	South Pacific Ocean Sheet I
621	South Pacific Ocean Sheet II
622	South Pacific Ocean Sheet III
21030	Cabo Velas to Punta Mala
21036	Gulfo Dulse to Bahia de Paita (Galápagos Planning Chart)
21621	Isla de Coca
22000	Archipelago de Colon
22521	San Cristobal
22522	Anchorages of the Galápagos Islands
22523	Approaches to Canal de Santa Fe San Cristobal, Barrington
22524	Isla Espanola (Hood) and Approaches
22525	Bahia Gardner and Approaches
22527	Bahia Post Office
22526	Isla Santa Maria and Approaches
22528	South Coast of Island Santa Cruz—Isla Santa Fe
22529	Isla Isabela to Isla Santa Maria
22531	South Approaches to Canal de Pinzon
22532	Puerto Villamil to Punta Cristobal
22533	West Approach to Bahia Elizabeth
22541	Bahia Elizabeth Anchorage (Isabela and Fernandina)
22542	West Approach to Canal Bolivar
22543	North Coast of Isla Isabela
22544	North Approach to Isla Isabela
22545	Approaches to Bahia Cartago

22546	Bahia Cartago (Isla Isabela)
22547	Eastern Appraoch to Canal de San Salvador
22548	Isla Baltra
22549	North Coast of Isla San Salvador
22551	Isla Pinta and Isla Marchena
83020	Marquesas Islands
83023	Rangiroa to Katiu
83207	Ile Nukuhiva
83218	Hiva Oa, Tahu Ata and Montane
83382	Tahiti and Moorea
83385	Papeete Harbor
83392	Ile Huahine to Ile Maripiti
83397	Bora Bora

Coast and Geodetic Survey

5020	San Diego to San Francisco Bay
5105	North San Diego Bay
5107	San Diego Harbor
5114	Anacapa Passage
5116	San Miguel Passage
5120	Point Hueneme to Santa Barbara
5142	San Pedro Channel
5202	Point Dume to Purisima
5302	Point Conception to Point Sur
5386	San Luis Obisbo Bay and Approaches
5387	Morro Bay
5402	Point Sur to San Francisco
5403	Monterey Bay
5476	Pfifer Point to Point Cypress

H.O.

1044	West Coast of Baja
1193	Bahia de San Quintin to Isla Cedros
1310	West Coast of Lower California Point Cedros to Abrejia
5184	Point Loma to Punta Colnett

Naval Oceanographic Office

528 San Francisco to Point Telmo

SPARE PARTS LIST CARRIED ON *PAGURUS II*

1 Alternator #2871713, Delco #110842—42 Amps.
1. Aqualarm Temp. Sending Unit—installed
8 Bolts for Injectors #0095315
1 Clip Kit for Alternator
2 Cylinder Head Joints #36855108
1 Delco Voltage Regulator #2872513 Delco #1119513 negative ground
6 Elements Fram #CC1133 PL
6 Elemts Fram #CS1133PL
1 Fuel Lift Pump #2641314
2 Fuel Elements #30242
1 Fuel Line (#1) #35355268
1 Fuel Line (#2) #35354265
1 Fuel Line (#3) #35354263
1 Fuel Line (#4) #35354264
4 Gaskets for Cylinder Head Exhaust Ports #0490656
2 Gates Belts AX38 #8300A
1 Hose (Water Cooling Assembly) #34821112
1 Hose for Water Cooling Assembly #34831503
1 Hose for Water Cooling Assembly #34833121
2 Injector Assembly #2645559 (2 replaced within the week before leaving)
2 Injector Flanges #36173452 (later found to be unnecessary)
1 Jabsco Water Pump (#35615) #3270
3 Nuts (Fuel Oil Filter) #0576111
2 Olives (Cylinder Head Cover and Rocker Arm Assembly) #0566002
1 Parts Manual #3091
4 Washers (Fuel Oil Filter) #0920146
1 Water Puppy #8360-0001
10 ft. 12 guage wire
6 Injector gaskets for base of injectors
40 Washers for fuel line leak off at injectors

1 Oil filter element #26540154
Left Sensitivity relay for autopilot #1000
Right Sensitivity relay for autopilot #1001
1 kit for repairing toilet